U. S. Fire Administration

Attacking the Violent Crime of Arson

A Report on America's Fire Investigation Units

October 2004

Homeland
Security

Table of Contents

Introduction... i

Section 1. The Impact of Arson: What National Data Reveal 1

Section 2. The U.S. Fire Administration Fire and Arson Investigation
Technical Assistance Project.. 5
Background 5
Methodology.. 6
Comments on the Project from the Field... 9

Section 3. Trends and Best Practices 13
Composition of Investigation Units 13
Strike Teams and Task Forces... 13
Professional Standards and Training.. 19
Prosecutor Support... 21
Investigation Data and Reporting 22
Evidence and Laboratories... 25
Use of Accelerant Detection Canines 25
Investigator Work Schedules 27
Unit Management 31
Caseload 36
Reaching Juveniles Who Set Fires 40
Firefighter Arsonists 42
Intelligence and Surveillance 43

Section 4. The Tennessee Interoperability Project--A Statewide, Multiagency Solution
to Data and Intelligence Sharing... 45

Minimum Essential Requirements for Computerized Case Management and
Intelligence Sharing System 46

Applications for Emergency Onscene Reporting and Interagency Communications 47

Appendix A
U.S. Fire Administration's Project Application 49

Appendix B
List of All Participating Jurisdictions 53

Appendix C
Copy of Ohio's Seminar Program Relative to Prosecution of Arson Cases 57

Appendix D

Sample of Work Schedule from Many Jurisdictions 58

Alaska State Fire Marshal's Office 1

Delaware State Fire Marshal's Office 1

Maine State Fire Marshal's Office 1

Maryland State Fire Marshal's Office 1

Massachusetts State Fire Marshal's Office 1

Rhode Island State Fire Marshal's Office 1

Utah State Fire Marshal's Office 2

Fairbanks, Alaska 2

Mesa, Arizona 2

Tucson, Arizona 2

Bakersfield, California 3

Fresno, California 3

Ft. Lauderdale, Florida 3

West Palm Beach County, Florida 4

Cobb County, Georgia 4

Fulton County, Georgia 4

Des Moines, Iowa 4

Honolulu, Hawaii 4

Kansas City, Kansas 5

Flint, Michigan 5

Oklahoma City, Oklahoma 5

El Paso, Texas 5

San Antonio, Texas 5

Salt Lake City, Utah 6

Virginia Beach, Virginia 6

PROJECT TEAM

U.S. Fire Administration

Kenneth Kuntz, Fire Studies Specialist

Bryan S. McCreary, Contract Officer

Gregory Blair, Contract Specialist

TriData Corporation

Hollis Stambaugh, Project Director

Joseph Ockershausen, Project Manager

Teresa Copping, Project Assistant

Lisa Aziz, Assistant

Other Investigation Management Specialists (1989-2004)

*Ward Caddington

*Richard Crispen

*Paul Flippin

*Randolph Kirby

*Richard McKee

*Stan Poole

*James Pott

*Paul Zipper

Daniel Carpenter

Robert Corry

Jane Edwards

Charles Jennings

Diane Martin

*Primary investigation specialists

Introduction

"How's a little fire, Scarecrow?" queried the Wicked Witch of the West before hurling a fireball at the frightened straw man in **The Wizard of Oz.** Using fire as a weapon is not just the stuff of movies, but a real-life ocurrence in communities across the United States. Commonly, the crime of arson is motivated by spite and revenge. Perpetrators strike with fire at buildings where people live, work, or socialize--causing injury, property loss, and death. Civilians and firefighters alike die in arson fires every year.

Thirty years ago, arson captured media attention because so-called arson-for-profit rings were burning down decaying urban neighborhoods that had ceased to be profitable, and then rebuilding them at a substantial profit. Other high-profile cases involved arsonists who were connected to gangs and drug lords, and who set fires to intimidate their rivals or as retribution for deals gone bad. Some of the most publicized cases occurred in the cities of New York, Boston, Houston, Los Angeles, Miami, Baltimore, and others. There even were situations where neighborhood vigilantes, who were frustrated with crime and run-down buildings, took it upon themselves to torch structures to rid the neighborhood of vagrants, prostitutes, and drug dealers.

Insurance companies were perceived as the main victims from intentional fires. As a crime committed against property, the economics of arson played center stage to the less well-defined statistics on injuries and deaths. Since arson fires do, on average, cause proportionately higher losses than fires from other causes, insurance companies committed many resources toward investigation and control. From establishing tip reward programs, training accelerant detection canines (ADC's), supporting arson reporting immunity legislation, and establishing the property insurance loss register (PILR), the insurance industry was a strong partner at that time.

There is a dichotomy between arson as a property crime and arson as a crime against people, and that lies at the heart of today's challenges with cases of arson. As a crime, arson's long-standing definition as the willful and malicious burning of property does not do justice to the fact that **today arson is usually a personal crime that is directed intentionally against specific victims.** It is time for arson to be dealt with as a violent crime against persons, not just a crime against property.

Today, spite and revenge dominate as the motives in intentional property fires, especially where there are casualties. Revenge-minded arsonists torch nightclubs, occupied residences, hotels, and other settings where their intended victims, and often other innocent people, are injured and killed. First responders get injured or die battling these blazes and trying to save others. Even though a portion of incendiary fires are motivated by other reasons (e.g., excitement, economic relief, peer pressure, a cry for help, and so forth) most set fires happen because someone wanted to inflict harm on another person using fire as the weapon of choice.

Fire investigation units from The U.S. Fire Administration's (USFA's) project indicated that spite and revenge were the most common motives behind incendiary fires. Among project sites from the past 5 years, spite and revenge ranked as the highest leading motives, when investigation units were queried about prevailing motives (see Figure 1).

Figure 1. Leading Motives by Frequency of Mention

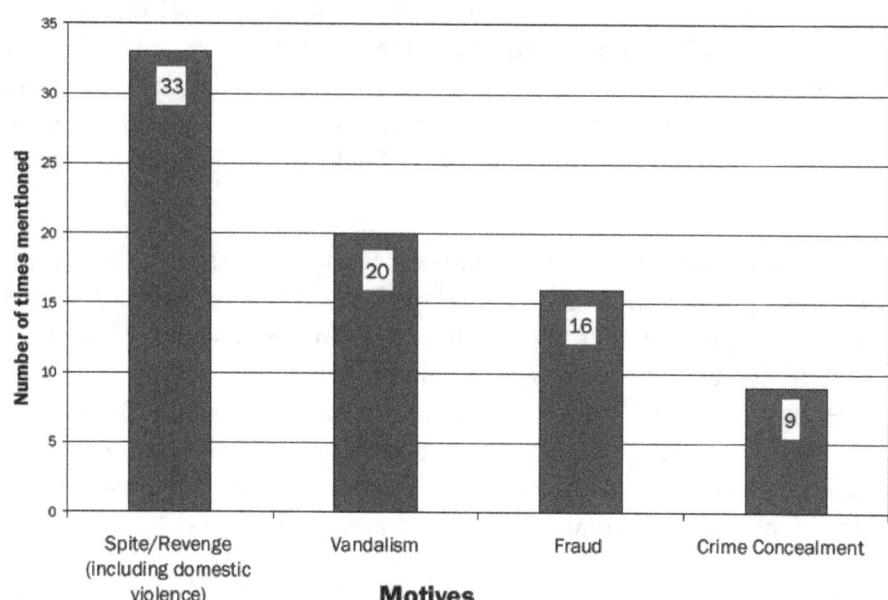

Criminal fires are the second leading cause of residential fires, and the number-one cause of non-residential structure fires. When firesetters employ accelerants to spread fire rapidly, or when they use multiple fire sets, the scene becomes even more dangerous. Responding personnel face high risks, sometimes paying with their lives. Over a 10-year period, from 1994 through 2003, there were 88 firefighter deaths in connection with incendiary and suspicious fires[1] -- almost 9 percent of all firefighter on-duty deaths. **That arson is a violent crime is apparent.** Funding that adequately supports investigations, training, and prosecution is essential.

The USFA supported State and local arson control units for 16 years, working to improve coordination among fire, police, and the court system and to build stronger fire investigation units. The Fire and Arson Investigation Technical Assistance Program served 143 State and local fire investigation units. It concentrated on facilitating multiagency cooperation and information sharing as well as clarifying operational roles.

This report compiles the best practices and common problems of fire protection and criminal justice agencies in identifying, investigating, prosecuting, and preventing arson. Trends, current challenges, and best practices are discussed. The first part of the report, Section 1 provides national data on arson, information that shows why the crime of arson needs the sustained attention of State and local elected, appointed, and public safety officials. Section 2 of this report describes USFA's technical assistance program and identifies the participating jurisdictions. Best practices are highlighted in Section 3, and in the last section, Section 4, a unique data management and intelligence-sharing project in Tennessee is showcased as a solution for the future.

---

[1] Fahy, R.F. and LeBlanc, P.R., *Full Report, Firefighter Fatalities in the United States-2003*, National Fire Protection Association, June 2004.

Section 1. The Impact of Arson: What National Data Reveal

The USFA maintains data on fire incidents by means of the National Fire Incident Reporting System (NFIRS). They also publish 10-year statistical overviews of the U.S. fire situation, the most recent of which is *Fire in the United States, 1989-1998, Twelfth Edition*. That report is based primarily on NFIRS data, but also includes National Fire Protection Association(NFPA) survey data, and information from the U.S. Bureau of the Census, the Consumer Price Index, the National Center for Health Statistics, and State Fire Marshal Offices.

According to *Fire in the United States*, in 1998 arson was the number-one cause of all fires (684,443 incidents) for all occupancies combined[2]. Almost 28 percent of fires that year were the result of arson. That was more than twice the second leading cause, open flame, which was responsible for 12.5 percent of the fires. Incendiary and suspicious fires were the second highest (19.7 percent) cause of fires involving deaths (1,684 cases); only smoking caused more fire deaths, and then only by a small margin (21.5 percent). Incendiary and suspicious fires caused more dollar loss than any other cause, and arson fires were second only to cooking-related fires in causing injuries.

There were 156,661 reported residential fires in 1998--the first year in which arson surpassed heating to become the second leading cause of residential fires following a dramatic downward trend in heating-related residential fires from 1989-1998. Arson ranked second as the cause of residential fire injuries and deaths and was the leading cause of dollar loss.

For nonresidential structure fires, arson became even more pronounced as a major causal factor. Table 1 reveals that arson outranks other causes in 7 out of 10 property types in a 3-year comparison.

Table 1. Comparison of Leading Causes of Nonresidential Structure Fires and Dollar Loss

Property type	Fires			Dollar Loss		
	1994	1996	1998	1994	1996	1998
Public Assembly	Arson	Arson	Arson	Arson	Arson	Arson
Eating, Drinking	Cooking	Cooking	Cooking	Arson	Cooking	Arson
Education	Arson	Arson	Arson	Arson	Arson	Arson
Institutions	Arson	Arson	Arson	Arson	Arson	Arson
Stores, Offices	Arson	Electrical Distribution	Electrical Distribution	Arson	Arson	Arson
Basic Industry	Electrical Distribution	Electrical Distribution	Other Equipment	Electrical Distribution	Other Equipment	Appliances
Manufacturing	Other Equipment	Other Equipment	Other Equipment	Other Equipment	Other Equipment	Other Equipment
Storage	Arson	Arson	Arson	Arson	Other Equipment	Arson
Vacant, Construction	Arson	Arson	Arson	Arson	Arson	Arson
Outside Structures, Unknown	Arson	Arson	Arson	Arson	Arson	Arson

Source: National Fire Data Center, USFA

---

[2] These rankings take into consideration a relative apportionment of unknown causes among all reported causes.

Various studies on fire trends emphasize the fact that over the last 25 years and since the widespread use of smoke detectors, the number of fire incidents has dropped substantially and continues to do so. Taking into consideration the fact that the U.S. population has grown over this time period, the rate of decrease is even more impressive. As a reflection of the overall decline in fire incidents, the number of intentional fires also shows significant reduction, from an estimated total of 860,000 in 1980 to 447,000 in 1998[3]. According to NFPA, the arson offense rate fell to 35.5 per 100,000 population in 2001--a four percent decrease over the previous year and an historic low.

As promising as the trends are, the United States still registers a fire death rate that is two to three times that of various European countries, and about 20 percent greater than the average for Europe as a whole[4]. Moreover, the numbers are more than digits, they represent events where people have suffered injuries, lost property, or even died.

In 1998, NFPA survey data showed that an estimated 2,705 civilians were injured in intentional fires. Applying ratios from the USFA's NFIRS counts produces an almost identical estimate of 2,712 civilians who were hurt in incendiary and suspicious fires that year. Intentional fires in 1998 killed between 674 and 682 civilians, according to these same sources. Almost all of these deaths occurred in intentional structure fires. It is heartbreaking enough when families suffer from fires that start accidentally; it is a terrible crime when people are injured or die from fires that someone purposefully sets.

A sense of the national arson picture also can be drawn from the Federal Bureau of Investigation's (FBI's) National Incident-Based Reporting System, or NIBRS, that is part of the Uniform Crime Report (UCR). The numbers in this system are lower than those in USFA's and NFPA's data programs, primarily because the FBI does not include fires that are suspicious, nor does it factor in as arson a percentage of the fires where the cause is listed as unknown. Only willfully or maliciously set fires that have been so classified through investigation are included. Another reason for the lower arson incidence counts in the FBI's records is that law enforcement agencies are not always involved in investigating arson. When fire officials handle the whole investigation, the incidents are not always reported to the FBI. The FBI's reporting system, nevertheless, does provide relevant information on the crime of arson, including statistics on arrests and trends.

Table 2. Crime Index Trends for Arson
Percent Change in Offenses

Years	Crime Index Overall	Arson Crime Index
1999/1998	- 6.8	- 3.6
2000/1999	- 0.2	+0.4
2001/2000	+2.1	+1.7
2002/2001	- 0.2	- 3.7

Source: FBI Uniform Crime Reports for 2002

---

[3] Hall, John Jr., *International Fires and Arson*, National Fire Protection Association, May 2003.

[4] U.S. Fire Administration, National Fire Data Center, *Fire in the United States*, Twelfth Edition, 2001.

What Table 2 shows is that from 1998 to 2000, the arson crime index fell only half as much as the index for all crimes, and then actually rose by a fraction. However, from 2000 through 2002, arson crime indexes posted more improvement than the overall crime index. Arson offenses increased from 2000 to 2001, but by a slightly lower percent than did crime overall. Between 2001 and 2002, arson offenses were down by almost four percent compared to two tenths of a percent for all crimes combined. Year-to-year fluctuations in all crime categories are common.

The average dollar loss per arson incident is about $11,000, including all property types covered in the FBI's definition of arson: dwellings, public buildings, motor vehicles, aircraft, and other property. NFPA reports that in 2002 there were an estimated 44,500 intentionally set (including suspicious and a portion of unknowns) structure fires and an estimated 41,000 vehicles that were set on fire intentionally. The property damage costs from these two types of property combined reached approximately $1.14 billion, or about $13,350 per incident, higher than the FBI's estimate.

According to the FBI, law enforcement agencies cleared about 16.0 percent of reported arson offenses in 2001 and 16.5 percent in 2002. These percentages are very consistent with the clearance rate for all property crime in 2001 and 2002: 16.2 and 16.5 percent respectively. The clearance rate for all Crime Index offenses together is higher, about 20 percent, and for the most serious crime, murder, is about 63 percent averaged over 2001 and 2002. Burglary crimes are the least frequently cleared offenses--just fewer than 13 percent on average.

One should take into consideration that clearance rates reflect, among other things, the degree of manpower assigned to solve the crime. Thus, as a property crime, arson cases are cleared at a rate equal to that for all property crimes combined. Violent crimes, by comparison, are cleared at nearly three times the rate of property crimes, but typically benefit from a greater commitment of resources per incident. If arson were viewed as a violent crime, instead of as a property crime, then the clearance rate would register as being quite low, indicating a need to apply more resources toward solving arson crimes.

Section 2. The U.S. Fire Administration Fire and Arson Investigation Technical Assistance Project

Background

Since 1989, the USFA has provided onsite assistance to State and local officials who are involved in investigating fires, identifying the ones that are set intentionally, determining who is responsible, and carrying the case forward through prosecution and sentencing. With intentional firesetting being the number-one cause of all fires combined, the Fire and Arson Investigation Technical Assistance Project has been a major and well-justified effort to affect a type of fire that concerns both fire protection and law enforcement agencies. Especially because arson is, at the same time, a fire and a crime, strategies for controlling it require an array of skills, including knowledge of building construction, fire science, criminal behavior, evidence collection and processing, case development, witness interviewing, courtroom procedures, and many more. To handle intentional fires, an investigator either needs to be trained and certified in all these areas, or designated fire and police personnel need to work together closely. Therein lie the challenges to managing arson, and a principal reason underlying USFA's special assistance.

In the decade before the USFA initiated the arson management technical assistance project, they had built and promoted the arson task force concept. USFA worked directly with the U.S. Conference of Mayors to organize 2-day regional workshops in cities around the country. Host city mayors, from Seattle to Houston to Louisville, invited fire and police chiefs, law enforcement and fire personnel, insurance industry representatives, Chamber of Commerce leaders, and prosecutors to examine the benefit of establishing formal task forces and early warning systems to predict and respond to arson fires. Arson hotlines were created, the PILR was developed, and arson reporting immunity legislation was passed--interagency information sharing and cooperation. Successful surveillance and neighborhood watch programs, juvenile firesetter counseling, and blight control ordinances were showcased as examples that cities everywhere could implement to reduce the crime of arson.

By the end of the 1980's, many jurisdictions had organized successful arson task forces, including San Francisco, Seattle, San Diego, Chicago, Rochester, Houston, and Philadelphia. However, some of these encountered problems with funding and political and policy changes over the years that made it difficult to maintain the same level of readiness and investigative quality.

Many cities across the country faced obstacles in dovetailing fire and police expertise, and in getting prosecutors to try cases largely based on circumstantial evidence. Police and fire agencies kept different records and worked different shifts. Drug crimes began to soar, siphoning off important police resources and leaving many arson task forces without adequate criminal investigation experience. Arson unit managers asked USFA to provide assistance in structuring and managing investigation units. There also was interest in identifying "best practices" for dealing with juvenile firesetters, which by that time had grown to represent about half of the known incendiary fire problem.

Methodology

The USFA initiated direct technical assistance and management reviews to fire investigation agencies, both local and State-based units in 1989. The project was part of USFA's ongoing effort to strengthen and enhance the work of fire investigation units across the country. USFA provided the assistance to help improve arrest and conviction rates and promote arson prevention. One objective was to highlight the positive features of a jurisdiction's arson control operations and then make selected "best practices" available for participants in future years to use as models. In this way, investigation units could share successful initiatives. The second objective was to identify problem areas and recommend options for resolving those concerns.

Through a competitive bid process, USFA selected an internationally recognized fire research and technical assistance vendor to develop a process through which investigation agencies could apply for assistance. The contractor also was tasked with evaluating local and State arson units and assisting them in identifying strategies for improving the identification of intentional fires and how those are processed, investigated, and prosecuted.

USFA adopted a technical assistance application process, and each year renewed the opportunity for police and fire-based investigation units to apply. Announcement letters were mailed to local public safety agencies and to State fire marshals and State police offices. The announcements also were posted on USFA's Web site and the application could be downloaded from there. A copy of the application form is provided in Appendix A.

Jurisdictions were informed that the technical assistance focused heavily on improving the working relationships among fire, police, prosecutors and other State or Federal agencies and that they would receive strategic recommendations for closer cooperation. Case prioritization and management, arson data collection and analysis, investigation reports, training, and time-of-day staffing coverage also were key components of the site reviews.

When evaluating the applications that jurisdictions submitted each year, USFA compared the submissions using the following guidelines:

1. Investigation caseload.

2. Composition of the fire investigation unit (fire and police or all fire).

3. Number of personnel.

4. Willingness of senior fire and police managers to use the technical assistance for advancing joint operations and commitment to improving outcomes.

5. Impact of arson on the community and special factors.

6. Evidence of efforts to prevent arson.

7. Description of problems and need for technical assistance.

8. Overall quality of the application.

Geographic location also was taken into consideration, though in any given year there might be a disproportionate number of applications from one State or from one region of the country. Over the course of 16 years, however, the geographic distribution of locations served represented a reasonable balance among States and regions. A list of the jurisdictions that participated and the year they were involved is presented in Appendix B. These sites also are mapped, as shown in Figure 2. Insofar as the types of jurisdictions are concerned, 60 percent were cities, 16 percent were counties, and 10 percent were States (State fire investigators cover investigations in smaller and unincorporated areas where there is no designated local fire investigation unit, as well as assist established local investigation agencies).

Project personnel evaluated the applications and forwarded recommendations to the USFA Project Officer, along with copies of all the applications. USFA made the final selection, and staff informed the sites of their selection. The field assessment and technical assistance was organized around three stages:

Stage One--Information collection and preparation. The information on the applications provided a basic idea of current fire and arson investigation operations problems that organizations were facing. To prepare for the fieldwork, however, staff asked the sites to assemble and send more detailed information, such as training records, investigation activity reports, sample investigation reports (closed cases), call-out procedures, and more. Team members studied the background information in preparation for the site visit.

Each jurisdiction also was asked to develop a list of individuals who held the positions earmarked for interview, and to augment that list with others they deemed valuable to the onsite review. A schedule of meetings and interview appointments was prepared and sent to the site so that the participating personnel could confirm when they were needed. The schedule made it possible to use onsite time productively, and organized the fieldwork.

Stage Two--Site visit and field work. USFA's fire investigation, arson control, and management experts spent approximately 3 days in each area. They conducted interviews; reviewed case files, reports, and other data; and visited laboratories, checked evidence lockers, and evaluated office space. Team members also identified what vehicles, tools, and other equipment the investigators had, and how operations and cases were managed. Shifts, call-out procedures, operation procedures, performance reviews, training, and communications all were checked out during the site work. Familiarization tours of arson-prone neighborhoods (except in the case of State unit reviews) provided additional perspective on how arson was affecting the community.

Team members typically met with the following individuals: the fire marshal, unit commander, chief of detectives, Bureau of Alcohol, Tobacco, Firearms and Explosives (ATF) office, investigators, operations chief, fire chief, city or county manager, training officer, adult and juvenile court judges, forensic laboratory director, housing and community development director, data management staff, prosecutor's office representative (usually an assistant district attorney), and more. In the case of State units, the unit commander often recommended a nearby prosecutor and several volunteer fire chiefs who the team met with during the site visit, since it was not possible to travel through the State and interview public safety personnel in each region.

LOCATION OF ARSON PROJECT SITES 1989 - 2004

There were opening and closeout meetings with key individuals so that the project could be introduced, and later, the preliminary findings presented, before the team returned home. Those meetings provided opportunities for all concerned to raise questions and obtain clarification on issues about the purpose of project and the credentials of the team members, the schedule, and key results from the review.

Stage Three--Preparation of draft and final report. After the site visit, the team completed a draft report with findings and recommendations. The jurisdiction was invited to review the draft and provide comments. The final report reflected the changes noted by the community or State. Participating sites completed an evaluation of the services they received, and USFA asked them to implement as many of the recommendations as were feasible.

Over the many years of the project, the span of review grew and the services increased from a focus on fire, police, and prosecutor cooperation to a broader assessment of quality of evidence, use of technologies, blight reduction strategies, juvenile intervention programs, and a more comprehensive analysis of training, work schedules, and case management. Jurisdictions found it was a valuable exercise to have an objective third party examine their operations and make recommendations. Often, the recommendations corroborated what unit commanders might have identified earlier as areas needing improvement. Just getting various public safety agencies together to identify roadblocks and consider changes contributed many times to increased cooperation and understanding which ultimately led to better case outcomes. Participation in USFA's project also brought fire and arson investigation to the attention of a community or State's top officials, lending greater visibility to the Fire Investigations Unit's (FIU's) work. In later communications with the team, many investigation unit personnel stated that the recommendations helped them obtain management and political support for much-needed changes.

During the fieldwork, team members interviewed officials from fire and law enforcement agencies, prosecutors' offices, courts, housing and community development departments, forensic laboratories, local ATF offices, information management specialists, support staff, and sometimes, schools. Other work focused on examining records, physical space, and evidence lockers--checking case files and supplemental reports, and compiling more data and statistics.

Comments on the Project from the Field

Evaluation was a standard part of the process. The research team asked each participating State or local agency to evaluate the technical services and the written field report wherein detailed findings and recommendations on arson control were provided. Often, participants included a letter that explained how the technical assistance made a difference and what recommendations were being implemented. The Chief of the Arson Division in the *Louisiana State Fire Marshal's Office*, noted, "...I must say that the documentation fully established the NEED, and justified projected implementation expenses based upon the obvious increase in efficiency and production that would result... The Final Report delineated these priorities in a straightforward manner, and will, I am certain, serve as guide for future Arson Division planning and procedural operations...in the past ten (10) months, the Division exceeded a 40% monthly clearance rate on four occasions...resulting in recent quarterly clearance rate of 33.3%, 31.7%, and 34%... This is remarkable, in light of the fact the previous quarterly reports had clearance rates of 12%, 15.6%, 11%, and 24%."

Seattle's Fire Investigation Unit Commander wrote, "The advice and guidance provided to us during the site visit, and in the draft report, have been tremendously helpful to us. It also contributed to moving our unit forward in a variety of areas, some that we may not have done otherwise. New lines of communications opened during the site visit and...I highly recommend your program to other units..."

In the **City and County of Denver,** the Chief of the Department took time to say that, "We agree with the recommendations in this well-written report, and plan to present it to the appropriate City officials for their consideration...Please accept my personal appreciation, that of the Fire Prevention and Investigation Division, and of the entire Denver Fire Department for the thorough audit that was conducted and the comprehensive summary of your findings."

Throughout the years, the project team discovered many units that were doing an excellent job, and quite a few that were accomplishing much despite funding cuts and growing caseloads. One of the most effective State investigation agencies is the *Fire and Explosion Investigative Section in the* **Massachusetts State Fire Marshal's Office.** The Section Commander at the time noted, "I would like to thank you for bringing the issues that we could improve upon to our attention. I am attaching our proposed solutions to the issues in an effort to address the problem areas... We hope these proposed solutions will ensure a more quality service...to the citizens of the Commonwealth."

Pierce County, Washington, was another jurisdiction that participated in 2002. The County Fire Marshal wrote to USFA saying, "We found the entire report helpful...we felt this was a very beneficial experience, well worth the investment of time. The results of this project will assist us in improving everything from fieldwork, to budgeting, to management."

Though most of the communities that were involved in USFA's project were urban counties, cities, and States, some smaller jurisdictions submitted convincing applications that cited a particular need for assistance, even if the annual caseload was not as high as in more populated areas. One such example was **Chester County, Pennsylvania,** where a series of fires challenged a loose affiliation of volunteer fire companies, local sheriff's department, and State police investigators.

Chester County was unique because it was the *Chester County Office of the District Attorney* that led the effort to solve arson crimes through better fire and law enforcement coordination. The District Attorney sought out the USFA project as a resource to help create an arson task force and to build a structure for the newly formed coalition. After the site work and the report, the District Attorney expressed his appreciation. "I have just completed my review of your Final Report detailing your findings and recommendations...I am very impressed with your work, it appears that you hit a home run...We appreciate the time, effort, and dedication...Your document will enhance and support our efforts."

Some fire investigation units carry responsibility for investigating both fires and explosions, and their investigators are trained as bomb technicians as well as fire investigators. The **Prince George's County, Maryland, Fire Department** previously was an example of a unit whose investigators wore two hats: fire investigator and bomb technician. In the year of USFA's management review, the Fire Chief committed to using the recommendations. "The final report that was generated from this project has provided our Department with the important information that has and will continue to be used to improve the effectiveness of our Fire Investigation Unit. Without this report, many goals and objectives for the (unit) may not have been achieved. This document will continue to serve as a reference for the future direction of the unit."

While USFA received more comments than can be included in this report, the quotes that were just mentioned, and the additional excerpts from evaluation letters that follow, serve to illustrate how valuable the project was to many individuals at the grassroots level who every day determine the cause of fires, make arrests, and develop successful cases for prosecution. Beyond generating more successful case outcomes, USFA's project also was intended to prevent fire setting and reduce arson losses by blight control legislation, pattern analysis, and juvenile intervention programs.

Among these participants, some were in the throes of establishing a new unit; others had long established task forces and strong programs, but they needed to upgrade procedures or recovery from budget cuts. In those cases, management efficiencies helped to offset the impact of funding losses. Still other agencies sought ideas on how to handle criminal investigations after other crimes drew away law enforcement support and case followup became seriously affected.

Following are excerpts from other letters:

- "During the past several years, we felt that we were at best becoming stagnant...by the time USFA's TriData team arrived...three of the four veteran fire investigators had resigned, and I was desperate for any assistance...As a result of the analysis and review, submitted in an excellent report, the Charlotte Fire Investigation Task Force is back on track and regaining the support of many Federal, State, and local officials. Due partially to the team's recommendations relating to shoring up relations with affiliated agencies, we were able to bring to a swift and successful closure our church burning case..."

  --Chief of Fire Investigations, Charlotte, North Carolina Fire Department

- "I...take this opportunity to advise you of the high level of competence and professionalism exhibited by the USFA team...I was most impressed by their knowledge. It was real-world and not some 'ivory tower' silliness...Many of their recommendations have already been implemented...This is a great program, and one that will allow this highly specialized and complex field of endeavors to become an art that will hopefully be recognized as a true profession in the near future."

  --Deputy Fire Commissioner, City of Chicago Fire Department

- "I believe [the report] will help us to improve our operation significantly. I want to reiterate my commitment to implementing as many of the recommendations as are possible..."

  --Fire Marshal, Portland, Oregon, Fire Department

- "...A study of this quality and magnitude should be put to good use for improvements and not allowed to gather dust."

  --Director of Public Safety, Cobb County, Georgia, Fire Department

- "...[the team's] insight and experience in arson investigations and investigation management has already caused us to look at several areas in a new light and reevaluate our policies and procedures. Although the final report of their findings is not yet available, we have already begun implementation of several of their early recommendations."

  --*Texas State Fire Marshal, Texas Commission on Fire Protection*

- "...The team's methodology was sincere, probing, thought-provoking, and quite productive. The wrap-up meeting already has us moving forward, assured that the final report will serve as a strong instrument in updating our operations and enhancing the quality and productivity level of our arson squad."

  --*Fire Marshal, Richmond, Virginia, Fire Department*

- "...We have already started implementing several of the recommendations suggested. The State Attorneys Office has also sent two additional attorneys to the Arson School for Prosecutors, taught by ATF...We have changed several of our operating guidelines, and are in the process of modifying others to follow suggestions made in the management review. Your team, their professionalism and dedication, were exceptional."

  --*District Commander, Palm Beach County, Florida, Fire and Rescue Department*

- "...The Fire Administration is to be congratulated for providing this excellent arson assistance program and staffing it with exceptional professionals such as those who came to Maine... Their preliminary recommendations demonstrated significant insight to the arson problem and the positive steps that we can address to help solve this issue in Maine."

  --*Maine State Fire Marshal, Maine State Fire Marshal's Office*

Section 3. Trends and Best Practices

The USFA Fire and Arson Investigation Technical Assistance project provided a unique opportunity to monitor how arson control practices have developed over a decade and a half. No other program has generated the degree of detailed information and first-hand exposure to State and local fire investigation operations and outcomes as has this long-running USFA program. Progress has been made in investigation techniques and protocols since the program began in 1989. Some of the most important trends are discussed next, along with examples of selected "best practices" from several jurisdictions.

It should be noted that, with so many sites involved in USFA's program from 1989-2004, it was not possible to highlight every noteworthy example of a job well done, or a new way of approaching arson control. Also, situations change and what was observed and recorded in 1991, for example, may or may not be the same in 2004. Thus, for this report, we drew upon more recent examples, reasoning that conditions were less likely to have changed significantly in the interim. Nonetheless, even a year can make a big difference if budgets are cut or policies are changed, so the examples cited were in place at the time the jurisdiction was involved in the project, but may or may not be the same today. Whether or not those examples remain exactly as described now is not the key issue anyway. What is important is that the solution solved a pressing problem at the time and is something other investigation agencies might find useful for their own purposes.

Composition of Investigation Units

Since arson is both a fire and a crime, it does not fit neatly into the exclusive domain of either the fire department or the police department. Historically, this has tended to create a problem. Expertise from both disciplines is essential in solving arson crimes. Communities are faced with the challenge of dovetailing these two public safety functions and managing their cooperation to pursue intentional fire cases.

Today's fire investigator must possess many skills. In fact, ideally, an investigator would have practical experience as a firefighter and as a police officer. The investigator would be knowledgeable about fire chemistry and behavior, construction methods and materials, and reaction of materials to fire exposure--including burn rate and heat release. Incident management investigation procedures and methods for securing and preserving evidence are essential capabilities, as are skills in questioning and interviewing, arrest, judicial procedure, criminal behavior, and psychological profiles. If an investigator also knows something about bookkeeping, forensic science, computers, and surveillance--so much the better. Since it is practically impossible for one person to possess all of these abilities and knowledge, and to remain current in these diverse fields, it is logical that a fire and police team represents the strongest capability for dealing with criminal fires.

Strike Teams and Task Forces

Fire and law enforcement have found numerous ways in which to marshal their resources to fight arson. The three most common organizational models are as follows:

*   **Independent, but coordinated**--The fire department handles origin and cause only. If a fire is determined to be incendiary or suspicious, the police department carries forward with the criminal investigation and files the case with the prosecutor's office. Generally, there is no particular police investigator assigned to arson cases, and any officer or detective can be given the case.

- **Integrated fire and police team**--Fire and police personnel work jointly and usually are cross-trained, each with the power of arrest. The police department members are specially assigned, but usually handle other case assignments as well, unless the arson caseload is substantial. With this type of arrangement, the unit members collaborate on cases, but report to their respective police or fire commanders. There are some joint fire and police units where all personnel function under a single command. That arrangement is the ideal, but is uncommon.

- **Fire department exclusively**--Fire department investigators are responsible for all phases of cause determination and criminal investigation. For this arrangement to work, fire investigators must be trained and certified as law enforcement officers and have the power of arrest.

In 1980, the USFA began promoting the concept of an arson task force. Working with the U.S. Conference of Mayors, USFA sponsored 10 regional workshops nationwide, and invited mayors, fire chiefs, police chiefs, prosecutors, and insurance industry representatives to a workshop that described multiagency and community-wide task forces that could aggressively confront arson and begin to affect its spread. USFA covered investigation protocols, early warning systems, and community task forces involving the media and the Chamber of Commerce, plus strategies for addressing the developing juvenile firesetter problem. Successful arson task force commanders in several big cities spoke about the effectiveness of combined police and fire investigators in solving fire crimes.

Throughout the 1980's, the number of arson task forces grew. At the end of the decade, USFA initiated this project, originally with the goal of showcasing some of the more successful units and distributing a report with examples and best practices for other fire departments. What was discovered, however, was that numerous previously notable task forces had lost personnel and were being dismantled. The drug scene in America and its myriad associated crimes had begun to claim the law enforcement task force members who were pulled back to cover drug-related crimes, gang violence, and homicides. Many of the units were left with a cadre of experienced origin and cause investigators, but no seasoned law enforcement personnel and no arrest powers.

Initially, some fire chiefs were reluctant to support training fire investigators in law enforcement as a solution to the loss of police staffing. Likewise, some police chiefs disliked the idea of fire investigators having peace officer status and weapons, and lobbied to keep the criminal investigation assignment in the realm of the police department, even if they were not regularly assigning police investigators to arson cases. Where those conditions prevailed, arson investigations and the pursuit of arsonists were essentially in limbo. At the same time, training budgets leveled off or were reduced, so even when a fire chief supported cross training for fire investigators, the funding for special training and for backfilling an investigator's position while being trained was scarce.

USFA's technical assistance to fire investigation units filled a need and addressed the gap in effectiveness that had ensued following the dissolution of many task forces. Project personnel worked with States, cities, and counties in finding ways to maintain the essential expertise for fighting arson, while dealing with the realities of reduced budgets and a rise in other crimes that negatively affected the composition and functioning of investigation units. There also were intact fire and police units that requested USFA's assistance to improve aspects of their operations and procedures.

One of the best examples of a strike team was found in **San Diego, California**. The Metro Arson Strike Team (MAST) was established in 1979. MAST is a multiagency fire investigation unit of 12 full-time personnel provided by the San Diego Police Department (4), San Diego Fire and Life Safety Service (6), and the local office of the ATF (2). Six additional fire and police personnel are assigned to function as relief investigators and to fill short-term vacancies.

The on-duty shift captain supervises MAST's fieldwork, including initial origin and cause determination. The captain collaborates with the police sergeant assigned to MAST who assigns cases to one of the three field detectives. Criminal investigations are governed by the police department's criminal investigation protocols. All of the investigators are certified investigators through either the International Association of Arson Investigator or the National Association of Arson Investigators, and the State of California. Moreover, personnel assigned to MAST are cross-trained in both origin and cause determination and as peace officers through the California Commission on Peace Officer Standards and Training (POST).

A Special Agent with the ATF, who is a Certified Explosives Specialist, works out of the MAST office. The agent primarily assists with complex incidents involving abortion clinics and churches, incidents occurring on Federal property, and bombing and explosive recovery incidents. The agent is MAST's Liaison to the U.S. Attorney's office as well. A Special Agent from the FBI functions as a liaison with the unit, attends training with the fire and explosives personnel, and responds to incidents when requested.

Since the unit also is responsible for handling reports of suspicious packages and situations involving explosives, six additional personnel are trained to render operations safe and provide postblast investigations, scene documentation, and evidence recovery. These six explosive device technicians (EDT's) who work with MAST are assigned to the same station. Minimum staffing requires that at least one EDT be on duty at all times. Another EDT is assigned to day work and handles administrative duties, while another is available "on-call" to ensure a two-person response to explosives-related incidents.

Completing the support for the unit are personnel from the police department's Forensic Science Section, Neighborhood Policing Areas, and the Training Division. San Diego's fire and explosives investigation model uses the expertise and technical resources of all its member agencies. Consolidating the fire and explosives determination and criminal investigation functions into one specialized unit with one set of operating guidelines has been quite successful in the City's arson and explosives mitigation efforts.

Charlotte, North Carolina, created the Charlotte Fire Investigation Task Force in 1985 with a mission to investigate all fires, bomb threats, explosions, and false calls for service within the corporate limits of the city. At the time of the site visit, the nucleus of the task force included the chief fire investigator, the police arson supervisor, fire investigators, arson investigators, and clerical support. Additional resources, technical services, and support were available from affiliated agencies such as the State Bureau of Investigation; the ATF; Mecklenburg County Building Standards; and the Charlotte Police Department crime lab. Both the police and fire investigators were available 24 hours a day, 7 days a week, either by scheduled shift or call back, and they operated out of the same office. At the time, the average annual case closure rate was 48 percent and almost 60 percent of the fires investigated were determined to be incendiary fires. Suppression officers-in-charge determined origin and cause for "routine" fires, thus enabling the investigators to focus adequate attention on fires of greater consequence, both accidental and incendiary.

Charlotte's Task Force was well managed. There were clear dispatch and notification criteria along with comprehensive SOP's that spelled out the duties and responsibilities of all directly participating agencies as well as those that indirectly impacted operations, such as the Fire Communication Center. They had a large database of incident, victim, and witness information that could be queried online, and each investigator provided a monthly report to the Chief Fire Investigator. The report listed total cases assigned, cases status, warrants and convictions, court status, and case closure and outcome data. The Task Force had an accelerant detection canine. There was an exceptional amount of outreach and coordination with other city agencies.

The Task Force also worked with Neighborhood Development, the city agency responsible for inner-city revitalization, to support after-school supervision options for "latch-key" kids. The goal was to reduce the juvenile arson problem by providing supervision in the afternoons. At the schools, investigators worked jointly with police School Resource Officers when there were fires. In cooperation with the prosecutor's office, students who were arrested for fire crimes in schools were charged and prosecuted criminally, regardless of the extent of damage. The most stringent penalties were sought. That policy reduced fires set by students to disrupt class and get attention. Also, the Task Force had a good working relationship with the Housing Authority. The Authority made sure that vacant structures were secured to prevent individuals from entering the unit and setting fires.

Finally, Charlotte's juvenile firesetter intervention program, called "Firebusters," was court-mandated. Failure to participate or lack of family support resulted in further juvenile court action.

Chicago, Illinois, was successful with a structure where fire investigators and police detectives worked separately, but coordinated at both the supervisory and the field levels. At the time project personnel visited Chicago, well-trained and managed fire investigators staffed shift investigator positions as well as day work schedules. The latter provided followup work on origin and cause when the 24-hour platoon investigators classified a fire as "undetermined." Day-shift personnel also handled major case investigation in concert with the Chicago Police Department Bomb and Arson Section, the ATF Chicago Arson Group, and other law enforcement agencies where long and detailed investigative efforts were required.

When fire investigators determined a fire to be incendiary (58 percent of the investigated fires were incendiary--about 900 per year), the case was immediately referred to the Bomb and Arson Section in the Chicago Police Department's Bureau of Investigative Services. Fire department involvement continued as technical support to the police and other public agencies, including the Medical Examiner's Office, State's Attorney's Office and others. Twenty-two detectives and seven explosive technicians comprised the police unit, which worked closely with the fire investigators. The Police Evidence Unit and Fire Photographic Unit supported both the investigators and the detectives in forensic crime-scene processing and photography. There was a heavy emphasis on cross training and certification and regular command-level meetings were held between the Chief Fire Investigator and the Commander of the Bomb and Arson Section.

Tacoma, Washington, is a city of just under 200,000 population that has an 11-member FIU under the command of a Deputy Chief, also the city's Fire Marshal. In the unit there are three captains (Deputy Fire Marshals), seven lieutenants (fire inspectors/investigators), and one police detective. One of the Deputy Fire Marshals serves as commander of the FIU. Six of the seven investigators are certified as fire and/or explosive investigators through one or more professional organizations. The investigators' duties include code enforcement responsibilities, so they are available part-time to investigate fires. The commander, an FIU followup investigator, and the police detective are full-time fire investigators. This arrangement is an excellent solution for a city of this size. The inspectors gain credentials and experience in investigation so that, in the event of a major incident, they are able to assist, already familiar with the working of the investigators, and trained for the work at the scene. The Union and the fire department have entered into a Memorandum of Understanding (MOU) that newly appointed officers in special assignments, including the Fire Prevention Bureau, make a 3-year commitment to their assignments.

What is especially noteworthy about Tacoma is the Fire Chief's strong commitment to fire safety. The Chief did not allow an overall funding decrease in the department's budget to have a negative effect on the Fire Prevention Bureau and the FIU, choosing to cut back in other areas instead. The Chief reasoned that by having an active prevention, code enforcement, and investigation effort, structures will be safer and fewer fire responses will be required.

Other highlights of Tacoma's investigation program include excellent case management, use of solvability factors, quality assurance measures, accountability for reports, a juvenile firesetter program, and an experienced prosecutor with considerable understanding and knowledge of arson. She also is the chairperson of the Region Four Fire Counsel, and was certified in the arson prosecution course sponsored by the ATF and USFA. The FIU has an excellent cause-determination record: fewer than four percent of cases are undetermined.

Finally, the City of Tacoma ranks fifth in the nation in vehicle theft, so the fire department established vehicle fire response guidelines that prioritize the response to vehicle fires. Vehicle fires that fall outside the categories listed are considered minor fires not requiring an investigator. Fire investigators are not requested for passenger vehicle fires that occur while the vehicle is being driven, for example. Otherwise, investigators could be tied up with a growing number of vehicle fire investigations and be spread too thin on structure fire investigations.

The Massachusetts State Fire and Explosive Investigation Unit is one of the best State units in the country. The unit is well positioned to handle both current and evolving threats because it oversees two related sections that work cooperatively on cases: the Fire Investigations section and the Hazardous Device section (Bomb Squad). There are four fire investigation teams located geographically throughout the State. Additionally, there are three teams made up of two certified bomb technicians, and one bomb technician supervisor strategically located statewide.

Both the fire investigators and the bomb technicians have K-9 programs. There are three accelerant canine teams deployed regionally (a fourth was anticipated shortly after the site visit in 1999), and three nationally certified explosive detection canine teams. Interestingly, the Fire and Explosive Investigation Unit is funded through both the Department of Fire Services (pays for operating expenses) and the Massachusetts State Police (funds for salaries, vehicles, clothing, and overtime expenses). The joint fire and law enforcement commitments begin at the top and are reflected through to field operations.

All personnel are specially trained Massachusetts State Police detectives who exhibit an *esprit de corps* under the supervision of an innovative and supportive management team. Each year, individuals are recognized for their dedication and outstanding performance in matters related to combating arson. One year, for example, a State prosecutor and a chemist from the crime laboratory were recognized. After the award ceremony, attendees discuss ways to improve the unit's operations.

Management also provides its personnel with state-of-the-art equipment and opportunities for continuing education along with an expectation of high performance. Some of the resources made available are take-home vehicles, all investigation and evidence collection tools and gear, video and digital camera capabilities, laptop computers, a research library, and links to other specialized units such as surveillance teams, forensic technicians, cadaver K-9's, and Aviation Wing.

In Massachusetts, prospective candidates for the unit must have completed at least 100 hours of fire investigation training, participated in at least 50 separate fire investigations, have 2 years' experience as a fire investigator, and pass the fire investigator certification examination. A majority of the unit's personnel have advanced degrees. A minimum 5-year commitment is required.

State law requires that the State Fire Marshal investigate all fires and explosions of a suspicious or undetermined nature; however, their response procedures include conditions not generally articulated in response protocols, such as the following:

- fires involving the destruction of public records;
- fires or explosions where hate crimes appear to be indicated, or fires in places of worship;
- displacement of several or more families, or loss of commercial property with a subsequent loss of employment; and
- confirmed arson fires that damage occupied buildings.

Other circumstances warranting the unit's response are standard, except that none is based on estimated dollar loss as a compelling factor. Investigators employ a witness-driven and team concept for investigating fires and explosions, a fact that has improved arson prosecution results significantly. Investigators apply the Incident Command System (ICS) protocols to all major incidents, and personnel are trained on how to operate within ICS. The Fire Marshal's Office created a video featuring ICS as used during a major incident in 1995: the Malden Mills fire.

The Massachusetts State Police operated their own forensic laboratory and maintains an in-house staff of scientists and criminalists. Moreover, the unit has a designated evidence officer who is responsible for managing the totality of evidence and the chain of custody.

Professional Standards and Training

During the years of this project, the level and availability of training for fire investigators generally has improved. Professional standards have been developed for fire investigators that are based on NFPA 1033 and 921, their equivalent, or better. The outcome is increased professionalism in fire investigation work and a momentum among investigators to acquire certification. Fire investigators must keep pace with the latest fire science research, investigation technology, safety standards, and case law. If they carry weapons, they must remain qualified in firearms. In recent years, practically every unit has reported having at least one or two Certified Fire Investigators (CFI's). Achieving that milestone has become part of training goals in many cities and States.

In contrast, in the 1980's it was not uncommon for a new investigator to work cases armed with little more than on-the-job training supplemented with an investigation course or two after employment. Now, investigators often pursue training even before joining the unit. Police officer training accompanies the basic fire investigation training in jurisdictions where fire investigators are certified peace officers as well. Some communities assign new investigators to auto fire cases until they gain experience and are ready to handle structure fire investigations. Other departments mandate a "ride-along" apprenticeship for a couple of months so first-time investigators gain experience working side by side with more experienced investigators.

Despite advances in the amount and type of training offered, and in standards that define the levels of knowledge and practical experience that investigators should strive for, funding cutbacks have caused shortages that affect training as well as other facets of unit operations. Unit commanders must lobby to receive an adequate share of what is available. Fire and arson investigation is a specialized field in the fire service. Just like paramedics and hazardous materials specialists, investigators must remain current with training and certification. All these special skill areas require basic training and continuing education, and the funding necessary to guarantee continued competency.

Another common problem was that units lacked training policies that set basic and advanced training and continuing education requirements. Project personnel would discover, for example, that investigators might have taken a lot of courses, but that the training was not tied to specific requirements for the unit as a whole or for the investigator in particular.

It has been suggested that private companies might be good resources to help finance special training. Fire investigation agencies could accept donations from the private sector and pool these contributions into a central scholarship fund. The scholarship fund could become a permanent vehicle for annual giving so that future participation by businesses would become routine. Benefactors could be acknowledged publicly through the press and at an awards dinner. For the businesses, a scholarship fund would be a reasonable community betterment tax advantage that would build the capacity of local public safety agencies to control arson, a problem in which the business community has a vested interest anyway.

The experience gained through this project points to several specific types of training that are most needed. Law enforcement training for fire investigators is a must, unless the jurisdiction has a joint arson team with police detectives who are assigned full-time to the unit. Even then, while not as critical as when the fire department must work both the origin and cause and criminal investigation phases, cross training is beneficial for both police and fire to gain a perspective of the other department's role. Of course, this means that police investigators should take basic fire origin and cause courses, too.

There were cases, however, where the police chief or the fire chief would not authorize investigators to be trained in the skills of the other department because they wanted to keep a clear separation between the two public safety missions. Where this reluctance was present, typically it was because the fire chief or the police chief were uncomfortable with fire investigators having power of arrest. There were some valid reasons for their position. One was concern that armed fire investigators would add to a community's liability and possible legal action in case deadly force was to used inappropriately. However, in the particular jurisdictions where this was cited as a reason, inspections officers and environmental officers were certified peace officers, so refusal to do the same for fire investigators appeared to be based more on politics than on liability worries.

Some investigation units obtained sponsorship for law enforcement training through the county sheriff or the State police. Law enforcement training is essential for fire investigators who handle the criminal portion of the investigation as well as determining cause. Developing adequate investigation capabilities is necessary to serve the public. Ultimately, the fire chief and the police chief are responsible for setting a cooperative tone between their departments and for facilitating whatever training is crucial to cooperative investigations.

At the State level, investigators cover large geographic areas usually comprised of small towns and rural communities. If State investigators do not have police powers, they must rely on county sheriffs or a small local police department (with maybe three or four personnel) to carry the investigation after the fire cause has been determined. Those offices should be involved, in any case, but the State investigator needs status as a peace officer since the training, ability, and willingness to participate in arson investigations vary greatly among police agencies in sparsely populated areas. State police are another option, but they tend to be spread thin already. In Vermont and Massachusetts, the State fire investigators operate under the direction of the State law enforcement agency. In general, State investigators need full criminal investigation authority so they are qualified to handle all aspects of incendiary fire cases that extend beyond local law enforcement and prosecution.

Another training need is for line firefighters to be updated periodically on arson awareness so they can contribute to an investigation and avoid actions that could destroy evidence or otherwise compromise processing the scene. The team also found that Incident Commanders (IC's) needed to review the circumstances under which they should be making the call on origin and cause so that the investigators could concentrate more effort on the incidents that clearly required their special training. On the other hand, some officers at the scene were not calling out the investigators often enough. Striking a balance between calling out the investigators often enough or too often is a challenge for almost every department. USFA encourages use of call-out guidelines that are based on the IC's establishing origin and cause whenever reasonable, and then summoning fire investigators according to pre-established criteria. Using this system gives investigators sufficient time to focus on more involved cases and prepare detailed reports that support prosecution.

Prosecutor Support

One of the biggest problems that investigators faced in the late 80's and early 90's was obtaining support from the prosecutor. Though there still is resistance to developing these cases in some locales (many prosecutors have little exposure to incendiary fire cases), there has been a shift toward closer cooperation over the years. One contributing factor has been the joint USFA and ATF course for prosecutors, which is offered several times a year in different regions of the country. Prosecutors who attend the 3-day seminar learn about the causes of fires and how they develop. Thermal dynamics heat release and flashover are demonstrated through live burns, and the trainer explains how fire investigation processes a scene, collects evidence, and evaluates accident causes. Even in cities and counties where the prosecutor has not been involved in arson prosecution training, we found that prosecutors were aware of the course and hoped to participate in the future. Prosecutors benefit from instruction on the physical nature of fire, flashover, collapse, and incendiary devices and explosions. Such training builds a greater understanding of criminal fires.

In October 1995, the Texas Commission on Fire Protection, Office of the State Fire Marshal published a terrific "Arson Prosecution Handbook" and made it available throughout the State. The Handbook covered the arson laws in Texas; motives; agencies involved in fire investigation; fire chemistry and behavior; investigative techniques; and evidence collection, scientific aids, and laboratory analysis. Though the publication is 10 years old, the topics and contents provide an example for other State Fire Marshals. Since many prosecutors cannot attend training courses on arson prosecution, a handbook with basic information and guidelines provides a primer with valuable information and reference.

The State of Ohio has made a major commitment to training the officers of the court. With strong support from the Division of the State Fire Marshal (DSFM), State legislators passed a law mandating that the DSFM conduct annual training seminars for judges and prosecutors on how to prosecute arson cases successfully. Under this innovative program, the legislation requires the fire marshal to cooperate with the State Attorney General in organizing the biannual training seminar. Each prosecuting attorney may attend, or require an assistant prosecuting attorney to attend. A copy of Ohio's legislation is included in Appendix C.

As the professionalism of investigators has increased and the awareness of prosecutors has grown, each group tends to view the other in a more favorable light than was true when the project began. During site visits over the past 6 years or so, prosecutors generally expressed satisfaction with the quality of cases presented to them by the investigators, while investigators more often viewed the prosecutor as an ally, not as a roadblock. This stands in contrast to earlier in the program when many prosecutors who were interviewed were not interested in arson.

The widely held opinion a decade ago was that arson cases were extremely difficult to win, except under extraordinary circumstances, and since losing cases brought down the win ratio, there was little motivation to invest much time and effort on criminal fires. Nevertheless, there were a few outstanding prosecutors who rallied to bring arsonists to justice, even building their careers on a tough-on-arson stance and successful prosecutions. Among the USFA project sites, there was a prosecutor from Chicago, one from Minneapolis, and another in San Francisco who served as excellent examples. In those cities, an aggressive prosecutor helped get arsonists off the street; boosted the morale of investigators who knew that if they had a good case, the prosecutor would accept it; and built an enviable reputation that eventually lead to a judgeship. One San Francisco Assistant District Attorney won guilty verdicts on several arson cases with convictions that carried life sentences.

Years ago, a prosecutor in Chicago helped to establish the Cook County Arson Prosecution Task Force. Personnel from that task force met jointly with detectives from the Police Bomb and Arson Section and investigators from the Office of Fire Investigation in pre-trial meetings. The pretrial conferences enabled the prosecutor to be well prepared for all pending trial presentation. Buttressing good preparation and aggressive prosecution in Cook County was an Illinois arson sentencing law that provided for the following:

- mandatory life (no parole) in multiple-death arson cases;
- six years (no parole) in any cases involving an inhabited structure; and
- no statute of limitations for prosecuting arson cases.

One possible reason for the trend in increased prosecutorial support could be that this crime finally is being seen as a violent crime--an act intended to hurt or kill people. Sometimes arson-related injuries and deaths are counted in the one's and two's--sometimes in the dozens, as was the case when an angry boyfriend poured gasoline into the entrance (and only escape route) of the Happyland Social Club in New York City in 1990. Eighty-seven people died. Other high-profile cases, such as the $12 million ecoterrorist burning of a Vail, Colorado, ski lodge, raised awareness about the serious effects of arson. That case was especially worrisome because the terrorists first set fire to the repeating station and tower to delay first responders from getting to the fire scene at the lodge soon thereafter. The Phoenix FIU--as part of a task force consisting of Federal, State, and local law enforcement agencies--successfully closed a serial arsonist case in which ecoterrorists were burning new $500,000 to $1 million homes to protest urban sprawl.

Investigation Data and Reporting

In the 1980's, the majority of fire investigators wrote their investigation reports by hand and submitted them to an administrative assistant who typed the reports and then returned them to the investigator for review. The secretary then typed the final version, and the unit supervisor reviewed it. The process was time-consuming and duplicated effort. Moreover, the ratio of support staff to investigators was almost never adequate, and as budgets tightened, administrative hours became even scarcer. Units with particularly high caseloads struggled to remain current on their reports. Also, it was not uncommon to find that investigators within the same unit used different formats, depending on how they preferred to document the results of their investigations.

The situation began changing early in the 1990's. Some FIU's were supplied with a computer or two that investigators shared for preparing reports. Typically, older investigators continued to write reports by hand, while newer investigators, who had the benefit of more exposure to computer technology, used the computer for reports. Now, most investigators use a computer. Laptops are gaining popularity for their transportability into the field, especially for State investigators who travel longer distances to remote locations.

Computer use has made a difference. For one, computers have promoted standardization in reporting, which in turn facilitates case management, prosecutorial review, recordkeeping, and information sharing. The best example of the latter is the system pioneered by the Tennessee State Fire Marshal's office in conjunction with the USFA. Section 4 provides details on that collaboration and what the results could indicate for the future.

In spite of advances in documenting information, and the unique example in Tennessee, investigation units have yet to take full advantage of the potential afforded by information management systems. The problem is a basic, long-standing one in the fire service--the usefulness of data has not been well articulated, nor has there been sufficient guidance on what outputs would be possible, or how to accomplish data mining. The first hurdle years ago was getting a standard national fire incident data system and obtaining buy-in from States and cities across the country. Today NFIRS is widely accepted and in use by almost all States and most fire departments. Similarly, the preponderance of investigation units now captures investigation-related data, though more needs to be done to establish a comprehensive, standard data set with such details as:

- incident profile (day, time, weather, conditions upon arrival, type of occupancy or vehicle, and so forth);
- owner/occupant;
- victims;
- insurer;
- names of witnesses and suspects (separate juveniles involved from adults), including inter views;
- origin and cause determination;
- arrest information, including age, race, prior arrest, involvement of drugs or alcohol in the crime;
- consideration of accidental causes;
- modus operandi;
- description of motive;
- photo log;
- evidence log;
- criminal investigation actions;
- prosecutor involved;
- warrants and arrests; and
- case outcome.

Over the past generation, significant strides have been made in gathering data. While that is noteworthy, it only partially solves the problem of adequate information systems for managing arson control. The data residing in systems are only as valuable as the extent to which they then are summarized, analyzed, and used to evaluate fire trends, evaluate program impacts, manage investigated cases, rate employee performance, track changes, develop budgets, and ascertain training needs. It is here where progress is stalled. Having the ability to transfer information electronically from agency to agency, from the field to headquarters, and from investigator to investigator does not in and of itself accomplish a great deal. Someone must take the initiative to link systems and enable users to share the emerging wealth of information. Management needs to support such efforts by endorsing technology along with the funds required to put improved systems into place.

There are several sources of expertise that departments could approach for assistance in developing investigation databases and report products. We found examples of these among the sites that participated in USFA's technical assistance project. Universities and community colleges often seek internships for students. This opportunity should be explored with the applied technology and information science divisions of local colleges. Increasingly, fire departments have their own information management assets, but such personnel are not always used to full advantage by the Fire Marshal's office to develop programs specific to investigation, inspection, and prevention requirements. Police departments may have a specialist who could lend a hand in setting up an information management program for fire investigators, especially if the community has a joint fire and law enforcement investigation unit where cooperation and information-sharing are already in place.

The crucial factor is that whoever the department selects needs to be experienced in systems integration, not just in designing a program. This was the key lesson learned from the Tennessee experience. Keep in mind that the goal is to maximize the usefulness and exchangeability of data within a select group of users, so computers have to be able to "talk" to each other in a shared environment that is at the same time secure.

Data related to investigations can play a pivotal role in the unit's success. The role of data is not immediately apparent, but is nevertheless as essential as the handtools used at the scene. For example, investigation units do not document motive in their records, probably because it has to be backed into the case file after a case is cleared and motive can be identified with certainty. The method used to set the fire is another detail that is not captured routinely. Yet, the reasons and methods involved in criminal firesetting are important for developing intervention, prevention, and surveillance activities.

Investigators usually know, or at least have a good idea of, what motives are behind the fires they investigate. At some point in the investigation process, suspected motives are either confirmed or altered based on more details about the suspect or acknowledged perpetrator. If that information is entered into the case database, then motive profiles can be produced that quantify the extent to which any given motive prevails and what the circumstances behind it are. Motive data can also be linked to the arsonist's age and be mapped against type of occupancy, vehicle, or outside fire. Virtually all the communities in the project see the full range of motives to one degree or another.

Anecdotally, outside and rubbish fires are usually the work of juveniles where boredom or a clear intent for criminal mischief prevails. Occupied dwellings or places of public assembly signal revenge as a likely motive. Keeping track of this information and cross-indexing it with data points from other cases will not necessarily reveal big surprises about motive. However, it will contribute to connecting the dots among multiple cases and monitoring trends, not only within the immediate jurisdiction, but also regionally among investigation units that may be dealing with similar profiles. Less has been documented about connections between modus operandi and motive, property type, or age of arsonist. That topic would be a valuable research endeavor for the future.

Finally, approaches for preventing the commission of arson can be tailored more accurately when motive, method, and other key information is documented and analyzed before committing resources. This information also serves as an essential baseline for later evaluating the relative successfulness of new and innovative arson control programs. Figure 3 on the following page provides an example of the detailed information investigation unit managers need from investigation personnel in order to manage cases and activities most effectively.

Evidence and Laboratories

The biggest changes in evidence collection and laboratory support observed over the years were the improvement in turnaround time for lab reports on fire debris analysis, and the growing use of accelerant detection canines (ADC's) to support scene processing. Also, more labs have acquired certification or are in the process of qualifying. The majority of units have secure facilities to store evidence and adequate tools. Canine teams are discussed in the next section.

One of the most common evidence-related problems seen was where investigators risked cross-contaminating the evidence supplies and samples. Some investigators stored evidence containers and tools in the trunk of their vehicles along with other items from the fire scene investigation. In evidence storage areas, another problem observed was situations where old samples dating back 10 years or more sat on shelves taking up space because there were no protocols for evidence disposal.

Use of Accelerant Detection Canines

ADC's were relatively new to the scene in 1989 when this project was initiated. A valuable forensic tool, ADC's have grown in importance to field investigators at both the State and local level. However, ADC's are not necessarily the right answer for every FIU. Moreover, some of the ADC programs the team reviewed were not providing a sufficient benefit for the cost involved. Generally, the reasons for that stemmed from the handler's laxness in the canine's regimen. For example, if the dog is allowed to receive treats from others, then the food-reward training and the dog's effectiveness are compromised. On one occasion, when the percentage of "hits" that were confirmed by the laboratories was below average, the problem actually was traced to the carbon strip in the evidence can; it was being placed in the lid by laboratory personnel after they opened the seal.

Standard procedures for collecting and preserving evidence, as with all cases, must be followed meticulously if investigators want to avoid negating the canine's work. Some K-9 handlers were not keeping records of the dog's hits and the laboratory's confirmation, which are needed for court.

A top consideration for any unit deciding whether to acquire an accelerant detection canine is cost in comparison to how much the dog really will be used. Though an unquestionably valuable resource, a department must be prepared to pay for the care, feeding, and veterinary expenses of the dog; the cost of the designated investigator who will be assigned as the handler; and the cost of replacing that investigator's time and regular caseload with an additional investigator. If the budget can handle that combination of expenses, and the annual caseload of suspicious and incendiary fires is big enough to support using the dog on a frequent basis, then a department should certainly consider adding an ADC to its unit. Sometimes the best solution for units in smaller cities or where there are not enough cases to warrant a dedicated ADC is to share this resource. In that way, the cost is split, but the canine is reasonably available and usually closer than would be the ADC from the State fire marshal's office or distant metro city.

Figure 2. Sample Format: Fire Investigator Activity Report					
Fire Investigator:			**Month/Year:**		
Activity	**Week 1**	**Week 2**	**Week 3**	**Week 4**	**Total**
Number of New Fires Investigated					
Structure Fires					
Vehicle Fires					
Outside Fires					
Number of Incendiary Fires					
Structure Fires					
Vehicle Fires					
Outside Fires					
Number of Accidental Fires					
Structure Fires					
Vehicle Fires					
Outside Fires					
Number of Unknown Causes					
Structure Fires					
Vehicle Fires					
Outside Fires					
Number of Civilian/FF Fatalities	/	/	/	/	/
Number of Civilian/FF Injuries	/	/	/	/	/
Incendiary Fire Dollar Loss					
Accidental Fire Dollar Loss					
Incendiary Fires Involving Adults					
Number of Adults Arrested					
Incendiary Fires Involving Juveniles					
Number of Juveniles Arrested					
Number of Cases Sent to Prosecutor					
Number of Cases Accepted					
Number Prosecuted by Trial					
Number Plea Bargained					
Number of Cases Null Processed					
Number of Cases Cleared					
Number of Cases Still Open					
Number of Postblast Investigations					
Other					

In units with a canine, the supervisor often volunteered to be the handler. Usually this does not work well because the supervisor can become so busy with the duties connected to the K-9 that management tasks, for example, case management, development of management information systems, task force development, and so forth, get short shrift. If at all possible, K-9 duty should be delegated to one of the other investigators.

Investigator Work Schedules

It was arguably the most sensitive issue of the project: the investigation work schedule. Project personnel encountered resistance when they indicated that a unit's work schedule needed to be changed to effect better results. It was that issue where the preferences of investigators, the needs of the community, and the realities of budgets tended to conflict the most. The teams observed every variation of schedule employed, examples of which are detailed in Appendix D, Examples of Work Schedules. The key factors that should drive whatever schedule a jurisdiction chooses are the days and times when most of the community's fires, and particularly incendiary fires, occur.

On a national basis, summary data from USFA's NFIRS for 2001 show that the most active time-frame for incendiary structure fires is late Saturday night to early Sunday morning (Saturday at 11 p.m. to Sunday at 4 a.m.), and then at 8 p.m. and 10 p.m. Sunday evening (see Table 3.) Outside fires set intentionally tend to occur from the afternoon to evening (between 2 p.m. and 9 p.m.) on any given day, but are more prevalent on Saturdays and Sundays. The same sort of table can be produced at the local and the State level once date and time information is entered into a database and analyzed.

The pattern is different for incendiary fires in mobile property. Table 4 shows that the intentional fires are set later at night and are more evenly distributed through the days of the week.

Not too surprisingly, intentionally set outdoor fires are heavily concentrated in the afternoons, a time when juveniles have free time and may be unsupervised. (See Table 5.)

Following are some of the factors that present conflicting demands for deciding what investigation schedule best serves the public.

1. The time period with the highest proportion of arson structure fires is 9 p.m. to 1 a.m.

2. Almost one-third of arson fires occur on Saturdays and Sundays.

3. Quick response to a fire scene is essential to interview bystanders and witnesses who can provide important information while it is fresh.

4. Followup work and coordination with other government agencies, places of business, witnesses, and suspects is usually accomplished during regular business hours, Monday through Friday.

5. Fair Labor Standards Act guidelines mandate overtime pay for working more than 40 hours per week.

6. Maintaining continuity and momentum on an investigation, especially during the first 48 hours, is critical, but difficult to do under certain shifts where an investigator may not see the case again for several days.

Table 3. Structure Incendiary/Suspicious Fires 2001								
Hour	Sun.	Mon.	Tue.	Wed.	Thur.	Fri.	Sat.	Total
Midnight	112	100	121	121	109	121	111	795
1 a.m.	108	132	93	101	107	117	133	791
2 a.m.	138	102	103	89	90	96	141	759
3 a.m.	125	85	84	89	88	99	107	677
4 a.m.	122	92	66	78	68	96	100	622
5 a.m.	79	58	59	73	58	60	66	453
6 a.m.	73	45	47	54	50	45	48	362
7 a.m.	60	53	49	41	46	48	45	342
8 a.m.	38	67	65	58	70	49	72	419
9 a.m.	57	71	89	64	73	63	54	471
10 a.m.	68	73	83	62	73	54	63	476
11 a.m.	72	97	97	85	78	79	71	579
Noon	80	90	86	91	103	97	96	643
1 p.m.	98	110	99	77	84	83	69	620
2 p.m.	93	79	98	102	71	100	97	640
3 p.m.	108	98	102	111	85	87	92	683
4 p.m.	110	99	95	97	108	101	103	713
5 p.m.	96	117	116	121	106	88	109	753
6 p.m.	119	114	127	104	90	102	101	757
7 p.m.	117	122	114	113	106	115	119	806
8 p.m.	130	111	102	122	94	92	110	761
9 p.m.	112	107	103	90	103	119	123	757
10 p.m.	134	128	109	98	112	108	127	816
11 p.m.	106	117	111	116	100	120	134	804
Total	2,355	2,267	2,218	2,157	2,072	2,139	2,291	15,499

Source: NFIRS

Table 4. Mobile Incendiary/Suspicious Fires 2001								
Hour	Sun.	Mon.	Tue.	Wed.	Thur.	Fri.	Sat.	Total
Midnight	187	198	168	175	139	146	162	1,175
1 a.m.	195	182	145	155	136	161	170	1,144
2 a.m.	186	154	136	129	110	150	157	1,022
3 a.m.	183	122	129	114	119	146	191	1,004
4 a.m.	160	100	93	94	94	119	148	808
5 a.m.	133	63	64	67	70	68	103	568
6 a.m.	83	43	34	40	37	44	53	334
7 a.m.	45	32	26	30	28	16	42	219
8 a.m.	27	31	19	20	18	18	27	160
9 a.m.	20	25	20	18	19	23	17	142
10 a.m.	30	28	18	30	26	16	22	170
11 a.m.	25	25	22	20	17	26	30	165
Noon	31	27	26	25	25	26	17	177
1 p.m.	28	29	24	35	30	26	26	198
2 p.m.	38	34	31	27	25	22	29	206
3 p.m.	37	34	28	32	32	34	36	233
4 p.m.	43	34	32	39	29	34	36	247
5 p.m.	46	39	31	44	29	29	32	250
6 p.m.	65	32	40	48	33	42	36	296
7 p.m.	85	46	51	50	43	63	76	414
8 p.m.	103	81	84	73	76	69	78	564
9 p.m.	121	100	93	96	96	93	117	716
10 p.m.	147	106	113	130	119	125	138	878
11 p.m.	171	160	151	149	122	155	145	1,053
Total	2,189	1,725	1,578	1,640	1,472	1,651	1,888	12,143

Source: NFIRS

Hour	Sun.	Mon.	Tue.	Wed.	Thur.	Fri.	Sat.	Total
			Table 5. Outdoor/Other Incendiary/Suspicious Fires 2001					
Midnight	208	165	129	129	175	171	190	1,167
1 a.m.	164	132	108	117	126	105	177	929
2 a.m.	165	91	94	97	110	81	168	806
3 a.m.	122	73	83	72	87	74	91	602
4 a.m.	87	52	63	47	51	54	74	428
5 a.m.	75	52	39	44	51	46	72	379
6 a.m.	49	46	45	50	56	49	55	350
7 a.m.	52	54	62	66	78	61	53	426
8 a.m.	60	67	70	79	86	85	77	524
9 a.m.	80	102	82	93	94	82	94	627
10 a.m.	109	125	111	134	106	118	158	861
11 a.m.	181	151	148	169	155	148	227	1,179
Noon	250	199	210	226	192	200	243	1,520
1 p.m.	322	260	251	232	235	214	345	1,859
2 p.m.	373	279	274	311	286	276	363	2,162
3 p.m.	385	364	330	306	329	287	370	2,371
4 p.m.	330	367	343	369	392	300	396	2,497
5 p.m.	391	448	390	393	362	307	404	2,695
6 p.m.	331	405	363	346	331	297	363	2,436
7 p.m.	332	357	321	351	299	325	365	2,350
8 p.m.	313	312	273	318	285	312	375	2,188
9 p.m.	279	290	265	309	252	281	326	2,002
10 p.m.	225	227	187	275	235	264	266	1,679
11 p.m.	168	182	159	218	172	200	251	1,350
Total	5,051	4,800	4,400	4,751	4,545	4,337	5,503	33,387

Source: NFIRS

7. Many departments have a cap on overtime costs.

8. A 40-hour, daytime schedule usually generates numerous evening and weekend callbacks and paid stand-by time. While the overtime pay earned is advantageous to the investigators, that situation usually leads to burnout and draws heavily on overtime dollars.

9. If fire investigators work in tandem with police investigators, the schedules need to be similar.

10. Changing the investigation schedule, whether from suppression shifts to a "regular" daytime schedule (8 to 5, Monday to Friday, or four 10-hour days), or vice versa, causes many objections and complaints to fire department managers. The first schedule is popular because it works easily with raising a family. Investigators who enjoy the longer stretches of days off that come with shift assignments, like the chance to work a second part-time job that augments their income.

A partial solution is to assign one or two investigators to day shift so the investigator going off duty could discuss whether the case needed follow up. It is recommended that the start time of investigators on 24-hour shifts be moved to noon, which would allow time for investigators to complete the reports from the previous night and to meet with the unit supervisor prior to going off duty.

State Fire Marshals are faced with an especially difficult problem regarding work schedules and burnout. Except in smaller States where even the most distant regions are only 2 hours away, State investigators average about 3 or 4 hours of driving time to fire scenes. That is a far different situation from a municipal investigator who usually can arrive within half an hour. At the time it was involved in the project, Nevada had five deputies to handle inspections and investigations in 14 large counties. With the office located in Carson City, investigators typically drove 8 hours to a scene, and then had to conduct the investigation, remain until the work was done, then drive back to Carson City. Fatigue was a real problem. The Wyoming State Fire Marshal's Office had a small caseload, but the two full-time investigators had to cover the entire state: 98,000 square miles.

The Florida and the Oklahoma State Fire Marshal Offices established regional offices, which helped reduce travel time and enabled investigators to arrive before evidence could be compromised. Oklahoma had two regional teams (Team East and Team West), which were further divided into districts. Supervisors and agents lived within their districts and operated out of home offices. Several State fire marshal offices use that configuration. It helps solve the availability problem, but presents special management and oversight challenges. In Florida there were seven regions. Each regional office had one or more field offices under their command, based on the size of the area served and the workload of that particular region. In Florida, State fire investigators must meet strict residency requirements. They must reside in their assigned county and be no more than 30 driving miles, nor more than 60 minutes drive time, of the duty post.

The Florida DSFM and the Bureau of Fire and Arson Investigations (BFAI) are dynamic, well-managed, and comparably well-funded organizations. The DSFM has a variety of statutorily mandated activities not generally associated with other State fire marshal offices. Operationally, the DSFM is divided into five primary functional areas, including the Bureau of Fire Prevention, Bureau of Standards and Training, Emergency Management and Response, BFAI, and the State Fire and Arson Laboratory. The BFAI is divided into two functional areas: field operations and special operations, which coordinates all Bureau training, criminal intelligence, research and development, communications, logistics, public relations, explosives ordnance disposal, technical rescue, and accelerant detection canines.

The BFAI strongly supports timely response to calls for service, as noted previously, but this is in danger of being compromised. The terrorist events of September 11, 2001, had a major ripple effect on the Bureau. Since tourism represents a large part of Florida's economy, and tourism declined significantly at the larger venues, tax revenues dropped significantly. In turn, the State reduced financial aid to cities and counties, so they are now demanding that the State fill in for the public services that cities have had to reduce. Fire investigation services are one example. According to one of the regional commanders, calls for State investigators in one major city increased from 60 a year to over 500, while another city stopped conducting both phases of the fire investigation process, relying instead on the State to assist. Florida's situation may be a harbinger of what is evolving in other States as well.

The challenge for unit commanders is to provide the best coverage for the community while living within the bounds of budget allocations and union rules. Because a large percentage of incendiary fires are set at night when darkness aids concealment of the crime, managers who have to hold the line on overtime are faced with getting a late start on investigating the fire by waiting until 8 a.m., or improving night and weekend coverage by scheduling investigators for regular duty during those prime times.

There is no perfect schedule, but there are some options that work better than others. In local FIU's, for example, schedules built around the three-platoon system correspond to the fire suppression work schedule and typically result in a 52-hour workweek. The advantage to this arrangement is that a trained investigator is available around-the-clock, Monday through Friday, which reduces reliance on callbacks and stand-bys. This shift guarantees quick arrival at the scene, regardless of day or time. The big disadvantage is that when the 24-hour shift is over, the investigator may be off for 72 hours during which time reports and followup on the fire investigation are on hold. Details that need to be worked immediately in the critical first 48 hours can be lost.

In contrast to the shift schedule, investigators working a 40-hour daytime schedule are in the best position to do the followup work and maintain momentum on the case. However, fires that occur during evenings and weekends must be covered by investigators who are called back to duty, and that incurs overtime costs. In any given week, the investigator with the callback rotation can burn out by midweek if the investigator "catches" a lot of fires in the first several days. It also is common for after-hours fires to be triaged more closely and held until morning, so as to avoid disturbing an investigator unnecessarily and to control overtime costs. This makes sense in terms of processing the scene, which is always more difficult at night, even with lights. However, the investigator loses the opportunity to access bystanders and witnesses immediately who could contribute vital information.

Unit Management

One of the primary purposes of USFA's project was to build the capacity of officials to manage fire investigation and arson control resources. Thus, management issues and tasks--such as case management and prioritization, information management, investigator supervision, liaison with other resource agencies, and related functions--were emphasized. In the process of working with the participating agencies, several management problems were found to be relatively commonplace. One problem is that managers tend not to have much time to lead or develop new initiatives with other agencies. Some managers define their role more as that of facilitator than as a commander. Some of the management functions that tend to receive less attention than they should are

1. Setting performance standards, and evaluating investigators on the basis of those standards.

2. Enforcing consequences for poor performance.

3. Restructuring work schedules to accommodate better coverage during evenings and weekends.

4. Updating and expanding data collection.

5. Applying data to guide management decisions on investigation priorities, prevention programs, and budget requirements.

6. Developing strong connections with key media and using the media to promote arson awareness and prevention messages, especially immediately following a major incendiary fire.

Some positive changes in management have developed over the course of the USFA project. Early in the program, many unit managers rotated into that position from outside the unit, and frequently had little real experience investigating fires or pursuing criminal cases. It was tough for them to do quality control on investigations or to offer consultation on procedures when they had not worked investigations themselves. Their role typically was that of caretaker and budget developer. The commander would head up the unit for 2 or 3 years, qualify for promotion or retirement, and move on to something else. Not surprisingly, there was little incentive for those managers to take on the difficult or unpopular parts of the job, especially if they were not going to be around long enough to see them through or to benefit from their implementation.

It is much more common now to see a unit commander who is the most experienced investigator in the unit, having been promoted from within. His or her task is to examine what the characteristics of the community's incendiary fire problem are; what mix of skills and programs are required to control arson; and how personnel should be assigned, trained, and scheduled. Stronger data and reporting need to be at the top of the list.

For FIU commanders, the ingredients for success are

1. Support from the top.

2. Adequate fire incident and investigation case data.

3. Flexible labor and management agreements (for recruitment and shift scheduling).

4. Funding for training and public education.

5. Good case management and method use solvability factors to prioritize cases.

6. Fire investigation and law enforcement experience.

7. Courage.

8. Creativity, commitment, and leadership.

Many fire investigation managers do have these attributes, but must operate within a political framework that discourages change. Fire and police chiefs should set the standard for cooperation and reward, not discourage efforts that accomplish operational and organizational improvements.

The most effective managers of the future will create arson prevention initiatives, organize community resources to combat arson, and advocate for juvenile firesetter intervention programs that will be tied directly to the court system where a judge mandates that juveniles and their parents participate. Investigators will establish close working partnerships with the prosecutor's office and with the homicide, juvenile, and domestic violence divisions of the police department. Intelligence will be shared. Arson control will be of concern to public agencies affected by arson, and they and community groups will support the unit's education and awareness efforts. Idealistic? Perhaps, but it is an irrefutable fact that fire departments and law enforcement agencies cannot solve the arson problem by themselves--any more than they can single-handedly put a stop to all crime or prevent all fires.

Experience with USFA's project reconfirmed that arson is rooted in a complex system of causes and conditions, all of which contribute to opportunities for this crime. Abandoned properties provide easy targets. Neglected and abused children adopt gangs as their families, protecting and defending that identity base through crime as necessary. Other children set fires out of boredom, delinquency, or desperation for attention. Parents are the ones who contribute to these conditions and behaviors and are responsible for the outcomes. Jealous lovers retaliate against their ex-partners by burning their residences or places of work. The Happyland nightclub fire illustrates how far the motives of spite and revenge can reach into the lives of innocent victims. Drug pushers stake out their territory and exact revenge using fire or explosives; murderers try to conceal their crime by burning the scene; and professional arsonists earn payoffs by destroying places where people earn a living or live.

A fire and arson investigation commander will be challenged to address all of these conditions, but a commander, with support from the top ranks of the fire department and police department, can take the lead in constructing a community-wide response. Tackling the labyrinth of conditions underlying arson (and all crime) is the rightful job of many stakeholders. Resources beyond those from police, fire, and the prosecutor are needed. Figure 4 provides a model of how a community can help. The model continues the emphasis on the primary roles played by the fire and police departments and the prosecutor, but expands the resources to stipulate support roles for such entities as the media, human service agencies, and others.

At the center of the model is data and information management. As mentioned previously, reliable numbers of the incidence of intentional (incendiary) fires, case statistics, characteristics of the arson problem, arrest data, and so forth are essential for any effective anti-arson effort. Leaders must identify and define the problem before they can develop strategies to solve it.

Figure 3. Community Partners in Arson Control

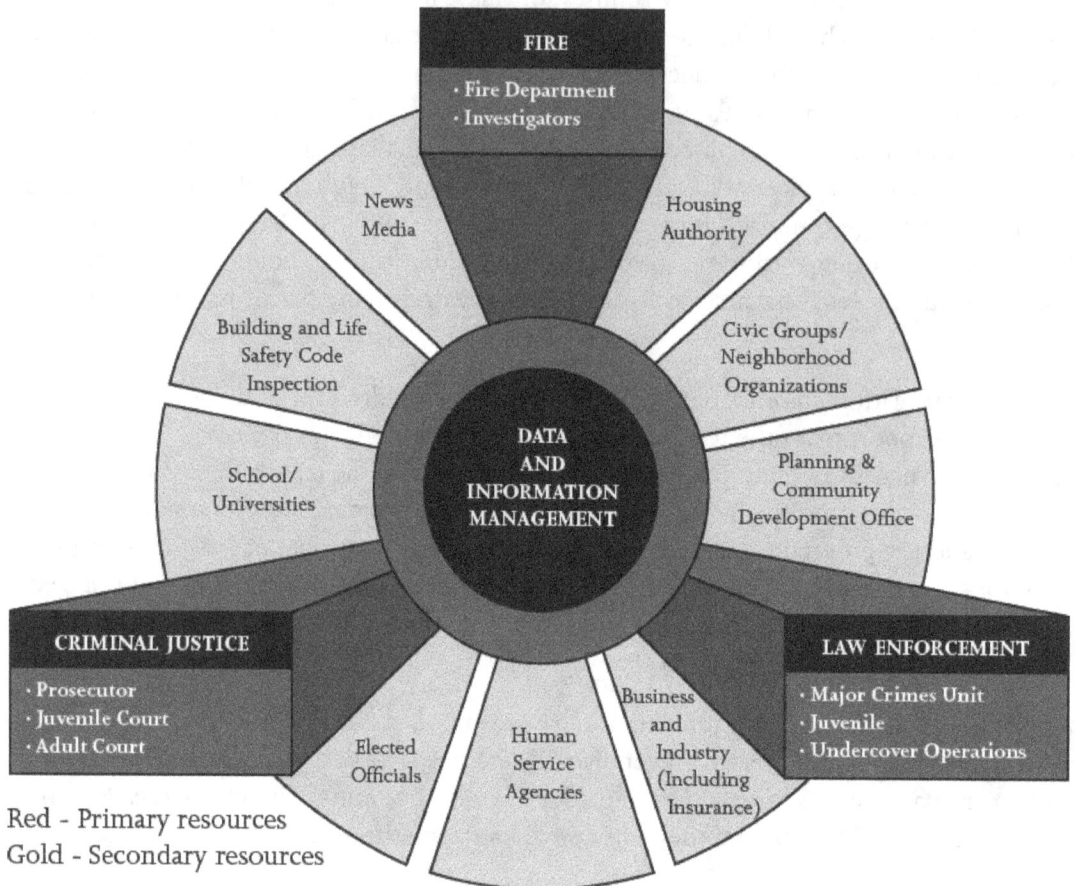

FIRE
· Fire Department
· Investigators

Housing
Authority

Civic Groups/
Neighborhood
Organizations

Planning &
Community
Development Office

News
Media

Building and Life
Safety Code
Inspection

School/
Universities

**DATA
AND
INFORMATION
MANAGEMENT**

CRIMINAL JUSTICE
· Prosecutor
· Juvenile Court
· Adult Court

LAW ENFORCEMENT
· Major Crimes Unit
· Juvenile
· Undercover Operations

Elected
Officials

Human
Service
Agencies

Business
and
Industry
(Including
Insurance)

Red - Primary resources
Gold - Secondary resources

Every community and State will determine for itself how many of the proposed supporting resources could be or should be tapped for assistance. A jurisdiction could organize the primary and secondary resources in any number of ways. For example, the players could be brought together under the umbrella of a community-wide arson control committee. Special subcommittees could address different aspects of the arson problem specifically: arson awareness and education, public relations and information, fundraising, legislation, housing rehabilitation programs, code enforcement, abatement of vacant structures, an ADC program, regional task force for major fires, and so forth. Some communities will prefer a less-structured solution. Regardless, the important point is that there are sources of help from outside the public safety agencies, and they can be asked to contribute to arson control.

Figure 4. Partners for Arson Control Initiatives	
PROSECUTOR	**LAW ENFORCEMENT**
• Train investigators in matters of evidence, courtroom procedures, and legal interpretation	• Open relevant law enforcement training courses to fire investigators
• Provide legal advice during investigations	• Provide surveillance and intelligence support
• Hold pretrial conferences	• Conduct criminal investigations
• Coordinate information on final case disposition	**FIRE**
• Participate in training for prosecutors	• Conduct origin and cause investigations
	• Conduct criminal investigations
	• Provide training
MEDIA (PRINT AND ELECTRONIC)	**PRINTERS/PUBLISHERS/PUBLIC RELATIONS AGENCIES**
• Provide air time/space to publicize arson prevention messages and programs	• Provide technical assistance in content, design, and layout of print material for arson control initiatives
• Provide technical assistance for development of arson/fire public education material	• Provide technical assistance in content and filming of training/educational videos
	• Print public education and awareness materials
COLLEGE/UNIVERSITIES	**DEPARTMENT OF PLANNING AND COMMUNITY DEVELOPMENT**
• Assign special arson related research through undergraduate or graduate programs in the criminal justice, sociology, social work, or government departments	• Assist in conducting risk analysis and neighborhood profiles in arson activity and losses
• Provide technical assistance to the fire department through the computer science department in establishing a fire incident and investigation data system and a data analysis protocol	• Link established neighborhood action committees to fire departments to obtain grassroots involvement
	• Investigate possibility of using some funding from Federal Housing and Community Development Program for arson control work
INSURERS	**BUDGET AND FINANCE OFFICE/GOVERNMENT LEGAL COUNSEL**
• Pool financial contributions each year to support a training scholarship for one or more local investigators to receive training through the National Fire Academy, the FBI Academy, State fire marshal's office, or International Association of Arson Investigators	• Review policy on disposition of assets seized during criminal investigations to allow fire and arson investigation divisions to benefit
• Contribute hardware, software, 35 mm cameras, video cameras, etc., to local or State fire investigation agencies	• Study possibility of initiating (or raising) fee collection for copies of fire investigation reports, files, and duplications of other frequently requested material
• Sponsor an arson hotline and reward program	• Support performance based/results oriented budgeting

Figure 4 outlines suggested activities for each entity, and the role they can play.

However structured, arson control groups should follow these recommendations to enhance the chance for success:

• Representatives on the committee should be senior-level people who have the authority to make decisions and commit resources.

- The members should be appointed formally by the local legislative body or chief executive for a fixed tenure.

- There should be a clear charter and mission.

- One of the three primary agencies should chair the group and guide deliberations. The role of chair could rotate among the three agencies.

An arson control group such as described above has the potential for contributing not only a pool of additional intellectual assets, but a host of valuable, tangible services as well. Once a group like this succeeds in its mission, the same sort of initiative can be used to deal with other far-reaching community problems. Oklahoma City, for example, developed a version of the community arson control committee years ago, which then tackled other neighborhood safety concerns as well.

To date few fire investigation managers have created community linkages and negotiated services and contributions from outside sources. The ability to broker such arrangements will be one of the defining attributes of managers in the future. They should be given the time to accomplish these sorts of ventures, which ultimately will pay dividends well beyond the cost of the commander's salary.

Caseload

What is a reasonable caseload for investigators? This was a common question among investigators, and the answer is not easily determined because it depends on many factors. Though none of the investigators indicated that their caseload was just right, and most believed it to be excessive, in fact the average annual caseload per investigator mostly ranged from being on the low side to being reasonable. Some units did carry higher than average caseloads on an ongoing basis, and a few units were so taxed that investigators could do little more than run from scene to scene with no time to develop cases and complete an adequate followup investigation. Typically, the teams found that the caseloads were acceptable, though monthly or seasonal fluctuations could cause sudden periodic surges that caused problems. It was not routine procedure to recommend more investigators unless the evidence clearly indicated the need for more personnel. In some cases, especially at the State level, justification for increased staffing was apparent.

The questions to consider when assessing caseload (too high, moderate, too low?) revolve around the issue of time--how much time it takes to get to the scene, conduct an examination, determine cause, identify a suspect, and develop a criminal case. Four issues in particular affect how much time an investigator is tied up with an investigation--and consequently, how many cases represent a manageable number.

- **Mission and range of responsibility**--For example, are the fire investigators responsible only for origin and cause, or do they handle the criminal investigation for incendiary fires as well? If the fire investigator's task ends when origin and cause are determined, then that unit can handle a higher caseload than would be acceptable for investigators who follow up leads, pursue suspects, and prepare cases for trial.

- **Travel time**—On average, how long does it take to get to the scene? State investigators (with the possible exceptions of Rhode Island, Hawaii, and Delaware) almost always have a longer drive to the scene than do local investigators. No investigation, however, comes close to the trek that Alaska's investigators routinely undergo. There, accessing a scene can require boarding small aircraft, flying over towering mountain ranges, and transferring to dog sleds.

- **Adults or Juveniles?**—Are adults or juveniles setting most of the fires? Cases where juveniles are involved typically do not take as long to solve.

- **Primary motive**—What is the most prevalent motive represented in the caseload? Fires set out of spite or as a corollary to domestic disputes tend to yield viable suspects more easily than fires set by a paid torch for economic reasons. Fires set to profit financially require an investigation into accounts, bank statements, and other financial information. Acquiring and analyzing financial records is time-consuming.

Carelessness is responsible for many accidental fires, so in order to prevent them, fire departments use a variety of good public fire prevention education programs to teach fire safety. Smoke detector programs, along with voluntary hazard identification home surveys and home escape planning, contribute significantly to eliminating fire hazards, and to survival if a fire does happen. The situation with intentionally set fires is actually not that much different.

Preventing incendiary fires requires initiatives that address both the targets and the perpetrators. Communities can reduce the targets of opportunity and stabilize neighborhoods by strictly enforcing codes and by upgrading the housing stock through housing rehabilitation grants, low-interest loans, and tax deferrals on improvements. Blighted structures that are beyond rehabilitation should be boarded up and removed as quickly as possible so they do not attract illicit activities. Neighborhood block watches help as do police monitoring of vagrants and gangs. Even weed control efforts and programs that improve neighborhood pride and appearance keep streets from inadvertently becoming arson-friendly.

There were three especially noteworthy arson prevention efforts that came to the team's attention during site work. Oklahoma City, Oklahoma, operated an arson/crime reduction task force when it was a project site, while Bridgeport, Connecticut, and Saginaw, Michigan, passed effective anti-blight ordinances. Memphis carried out a wide array of prevention initiatives.

Oklahoma City, Oklahoma: Arson and Crime Reduction Task Force—In one of the city's neighborhoods, escalating arson-related losses were the catalyst for a collaborative effort between neighborhood activists and city officials. Called "Project Impact" the program included the fire and police departments, the Inspection Services Division, the Neighborhood and Community Services Office, and neighborhood associations. In one year alone, the neighborhood, which was centrally located near the downtown district, had experienced 85 arson fires, about half of them in a single 3-month period. The area had a high crime rate, neighborhood blight and code violations, drug and gang-related activity, and a large transient population. A large number of vacant, unsecured, and dilapidated structures (residential and commercial) and many absentee landlords added to the host of problems.

The task force assigned specific actions to all the participating agencies. Neighborhood and Community Services conducted a sweep of the area and identified more than 200 structures that were abandoned or seriously code deficient. Violation notices were posted. If the violations were not corrected within the time allotted, the notice was submitted to city council, which issued contracts for securing or demolishing the premises. Neighborhood and Community Services also produced and distributed a pamphlet to all neighborhood residents that explained the task force's activities and encouraged the citizens to report problems.

The police department increased their activities in the area. They accumulated valuable intelligence by interviewing individuals who were found roaming after dark, and the information was shared with the fire department. Some of these individuals were potential suspects in the fires.

The fire investigation office worked with a confidential informant who infiltrated the vagrant and criminal parts of the community, identifying those who were responsible for the arson fires. The investigators then conducted surveillance operations, made arrests, and ultimately obtained convictions. Three arsonists actually were caught in the act of setting fires.

The fire and police departments used incident and investigation data to ascertain certain patterns and similarities in the fires that plagued the neighborhood. They mapped these to the residences of known suspects. Eventually, as these prime suspects were arrested and jailed, the fires in their respective areas decreased dramatically.

Fire and police personnel also increased the visible profile of marked vehicles in the neighborhood, thereby promoting citizen confidence in the city's commitment to improving the area. The increased visibility also discouraged other criminal activity. The final step was to encourage residents to reclaim the area by re-establishing neighborhood watch groups that had long been dormant. With assistance from the Neighborhood Alliance, these associations were contacted and began meeting. The residents again became the eyes and ears of the fire department, and displayed a renewed optimism in their neighborhood.

As a result of the task force, fire and police department responses to the area were dramatically reduced, and arson fires dropped 65 percent in a relatively short time.

Bridgeport, Connecticut: Antiblight Ordinance--The real estate boom of the 1980's made a significant impact on Bridgeport, Connecticut. Its proximity to New York City, its broad commercial and industrial base, and the easy commute to other Connecticut centers of insurance, banking, and defense manufacturing combined to make Bridgeport a hotbed of real estate speculation. At one point, it was estimated that there were three buyers for every piece of property on the market. Then came the bust. Expensive real estate fell in market value to prices below the amount owed on the mortgage. Absentee landlords were unable to find suitable tenants because employment opportunities in the area had declined. Property owners walked away from their investments leaving a proliferation of abandoned buildings. Fires and "urban mining" began to swell, a scenario that was repeated in many Rust-Belt cities at the time.

To combat the vacant and unsafe structure problem, the city passed an antiblight ordinance, which spelled out procedures for handling severely code-deficient properties. Owners of noncompliant properties were ordered to repair their properties and make them safe and secure, or to demolish them. Once their properties were identified as blighted, owners were subject to a daily fine until the structure was brought into compliance.

High profile political support was essential for the ordinance to be effective, so a special Vacant Structures Committee was created to enforce the antiblight laws. This standing committee included private and public sector representatives who met regularly to identify blighted buildings, act on cases, and initiate the process for rehabilitation or demolition.

Saginaw, Michigan: Teaming with the Water Department--In this innovative venture, Saginaw's fire department worked with the water meter program to obtain early indicators of property abandonment. The Water Department took regular, quarterly meter readings. The readings were computerized. The fire department arranged to have the meter reader push a certain key on the hand-held computer whenever the reader suspected that the building was vacant. Since vacant buildings comprised a significant portion of the arson problem in that city, the fire department could follow up with boarding and securing the buildings before they deteriorated, or to investigate whether the Dangerous Building Ordinance process needed to be initiated.

Memphis, Tennessee: Creation of Several Innovative Prevention Efforts--Among the various efforts to prevent fire and fire setting, the team identified the following programs during the field work in Memphis:

- Collaboration among the Division of Fire Services, the Memphis Housing Authority (MHA), and Memphis State University to develop and use fire experience data for a study and comparison of fires in all 22 MHA developments, including deaths and injuries, dollar loss, and causes. The report, "Analysis of Fires in Memphis Housing Authority Units" served as an important basis for subsequent actions.

- More rapid condemnation procedures were established when the city transferred condemnation authority from the city council to the Director of Housing and Community Development. Also the city could demolish the structure and attach the cost of demolition to the owner's tax bill.

- Inspectors representing housing, health, fire, and code enforcement would operate as a four-person team that jointly inspected problem occupancies. Dubbed "environmental teams," the groups learned about each other's areas of specialty, thus developing crosstrained personnel over time. The inspectors coordinated on court dates to present their collective evidence about deficient properties--a strategy that netted tougher action by the court. Community associations and the mayor's Citizen's Service Center contributed referrals about unsafe buildings. Correcting code violations materially improved the built environment and made it less vulnerable to both incendiary and accidental fires.

Reaching Juveniles Who Set Fires

Persons under the age of 18 set almost half the arson fires in this country. Crime Index statistics from 2001 for all crimes showed that nationally, 18.6 percent of the total clearances involved individuals under the age of 18. For arson specifically, the percentage of the clearances that involved juveniles jumped to 45.5 percent. The Uniform Crime Reporting (UCR) Program records as juvenile crimes cleared by arrest, those incidents where an offender under the age of 18 is cited to appear in juvenile court or before other juvenile authorities, even if the juvenile has not been physically arrested. Clearly, these are not incidents where curious kids are playing with matches, though that is another fire problem in and of itself.

The USFA produced a special report, *Arson and Juveniles: Responding to the Violence* (Report # 095), in their Technical Report Series. The report reviewed teen firesetting and interventions and included an examination of 35 incendiary juvenile-set fires provided by eight fire departments. While the number of cases was too small to be considered as representative of juveniles and incendiarism in general, the cases did present some interesting results. Nine (26 percent) involved occupied dwellings and schools. Another 13 of the fires (37 percent) were set in abandoned structures--the same as the number of outdoor fires that were set. Since the cases were hand selected against specific criteria--that there was a serious consequence to the firesetting--there was a built-in bias toward occupied structures. Nevertheless, when it comes to intentionally setting fires, juveniles are responsible for more than vandalism fires or criminal mischief.

When motive is factored in, a pattern tends to emerge both from the study cases and from local fire investigators' experience dealing with juvenile-set fires. Juveniles who, consciously or unconsciously, set fires to bring attention to difficult family circumstances are more likely to target occupied structures like their homes or schools. Situations where there are many family problems strongly indicate the likelihood of recidivism. Gang-related and revenge fires set by youth, on the other hand, occur more often in abandoned buildings (often used as drug houses or places to meet), but rarely in the offenders' own homes. These firesetters tend to set fires as a group, and often have a history of involvement with the juvenile justice system.

The significant involvement of youth in incendiary fires is striking, and the numbers probably are conservative due to the fact that many fires never come to the attention of the fire service. All fires set by juveniles need to be taken seriously. The size of the fire and the amount of damage are not good indicators of risk. Very often, juveniles who set fires start with small insignificant fires, then graduate to bigger fires. Many fire investigators know that they need to address today's small fires as though they could become tomorrow's fatal, multiple-alarm fires. Compounding the issue is the reported rise in juvenile bomb making. In Maryland, for example, a high school student brought two pipe bombs to school the second week after classes began in August 2004.

The Massachusetts Coalition for Juvenile Firesetter Intervention Programs tracked news stories featuring juveniles who were responsible for making bombs and setting fires. There were a total of 374 incidents captured over a 3-month period. These cases produced valuable data and insight into the seriousness of those incidents. Among the 374 cases, the following details emerged

- 155 fires;
- 95 bombs;

- 124 bomb threats;
- 40 fatalities;
- 159 injuries (including 6 responders); and
- 161 schools (43 percent) were targeted; houses were second (92 incidents).

In 1995, UCR reported that 52 percent of arson arrests and clearances were of children up to age 18. That percentage dropped to 45.5 percent in 2001, as mentioned earlier. Fortunately, juvenile firesetter intervention programs are in wider use today. A recent article in *The Strike Zone*, a newsletter published by the Massachusetts Coalition for Juvenile Firesetter Intervention Programs, cites an "estimated 4,000 interagency juvenile firesetter programs in our nation."

Most of the fire investigation agencies in this project claimed a juvenile firesetter program. However, the programs in some cases were informal and sporadic, e.g., talks at the fire station with kids known to have set fires. Few of the programs collected data and formally tracked the rate of recidivism over more than a few months--a year at the most. Other fire departments had developed excellent intervention programs that draw upon a variety of community resources and are tied to the court system. The latter makes a huge difference in program effectiveness because, when a judge mandates participation and the youth knows there will be consequences for noncompliance, attendance at the classes is much higher than when cooperation is voluntary. Examples of several good initiatives are provided next.

Cobb County, Georgia. The fire department and juvenile court in Cobb County developed a program for youth up to 18 years of age that aims to prevent future offenses by targeting elementary- and middle school-aged children who are first-time firesetters. Their mediation program was one of the first court-affiliated intervention programs for youth in the country, and reflected an excellent example of the concept referred to as "balanced and restorative justice." The primary goal is to get offending youths to take responsibility and be accountable for their actions, and then to change their behavior. Juvenile offenders had to meet with their victims and negotiate an acceptable restitution. There were both teen and adult mediators, trained by the University of Georgia. Acceptable forms of restitution included monetary compensation, yard/house work, services rendered to a retailer, or community service. The mediation agreements were binding.

Phoenix, Arizona. The Phoenix Fire Department established an excellent Youth Firesetter Intervention Program in 1980 that often has been used as a model for other fire departments throughout the country. The program was administered through the Community Outreach Section of the Urban Services Division. A professional with advanced degrees in education managed the program. Three components--education, mental health, and diversion--addressed specific aspects of firesetting behavior. The fire department assembled certified mental health providers to counsel youth firesetters and their families. If the family's insurance did not provide coverage for mental health services, there was a fund available to cover up to seven sessions.

The Maricopa County Attorney's Office reviewed the severity of all cases in which a juvenile had been criminally charged and determined whether they were eligible for the diversion program. Offenders in cases of a severe nature did not qualify. Phoenix maintained some data on its program, though statistics on recidivism were based on a 6-month followup evaluation postcard that relied on parents to report voluntarily any successive firesetting problems. Their data over four years (1996-1999) showed the following cumulative totals:

- over $4 million in dollar loss related to juvenile-set fires;
- almost 1,500 juveniles (includes noncriminal youth fire setting) referred by families, fire-fighters, and fire investigators; and
- 60 juvenile arson arrests.

Flint, Michigan. At the time of Flint's involvement in the project, the fire department had a juvenile firesetter program that screened and counseled child firesetters. They were in the process of tying into the Family Court Unit of the Genesee County Prosecutor's Office, which was responsible for juvenile offenders and family-related crimes. The Family County Unit had a full-time social worker and diversion programs supported by grants. These programs included "Get Scared"--a juvenile jail tour program--drug testing of juvenile offenders, and community service work for the poor. The programs were targeted at changing the destructive path that many of the city's juveniles were choosing.

The Family Court Unit was tough on violent or repeat juvenile offenders who were poised to become career criminals. One option was to delay sentencing for violent juveniles, a measure similar to probation before judgment, with the exception that delayed sentencing allowed the court to sentence the offender when the juvenile became an adult (for the crime committed as a juvenile). Typically, this occurred when the juvenile violated the provisions of probation, and continued to commit crimes.

Firefighter Arsonists

The unthinkable sometimes happens. A firefighter intentionally starts a fire. The incidence of illegal firesetting among the Nation's fire and rescue personnel is not known, and little research has been conducted on the problem. Most fire departments will never experience having a member indicted for arson. For those that do, however, the impact is significant. As in all incendiary fires, people can be killed or seriously injured. Homes and places of business can be destroyed. An arsonist from within the fire department can disgrace the whole department, and their actions diminish public trust.

As part of their Technical Report Series, the USFA produced *Firefighter Arson: Special Report* in December 2002. Researchers for the report documented case studies of firefighter arson. They also studied the impact of firefighter arson and presented information about several State and local fire service efforts designed to prevent criminal firesetting. Two agencies, the South Carolina Forestry Commission and the FBI's Behavior Analysis Unit have profiled the firefighter arsonist and their findings are compared in the report.

Generally speaking, firefighters who commit arson are white males, ages 17 to 26. Typically, family life was unstable or dysfunctional, and the individual lacks social and interpersonal skills. For these individuals, the appeal of being a firefighter stems from a sense of excitement about fire and a need to be a hero (often, they report or help extinguish the fires they set). These arsonists usually have average to above-average intelligence, though may have had only fair academic performance in school.

When a member of their own department sets a fire, fire investigators may have a harder time pinpointing who is responsible, because it is counterintuitive to suspect fire suppression personnel as arsonists. The situation is even more difficult when a fire *investigator* is the perpetrator, as in the case of John Orr, the former chief arson investigator in Glendale, California, who was a prolific arsonist and skilled at hiding his actions.

Intelligence and Surveillance

It is difficult to make an absolute determination about the trends in intelligence gathering and surveillance, but in general, surveillance operations are not carried out as regularly as they were 10 or 15 years ago. There is less interest in stakeouts, and investigators interviewed during the later years of the project reported they did not survey neighborhoods with unmarked vehicles very often. Personnel shortages may be a factor.

Intelligence sharing, however, appears to be on the increase, in large measure because better use of technology is facilitating information management. Florida and Tennessee, for example, have model programs in intelligence and data sharing. Florida has established a unique criminal intelligence gathering section to assist field investigators. There are seven crime intelligence analysts and one supervisor. One analyst is assigned in each regional office (seven offices). The analysts work directly on criminal information and intelligence related to arson investigations, and they provide statistical data to the managers and supervisors that is used to detect and prevent arson and related crimes. Intelligence unit personnel are distributed among the State's seven regional offices where they provide criminal histories and backgrounds on individuals and suspects. They conduct courthouse checks, assist with case management on major investigations, and do crime mapping of suspicious and incendiary fires. Staff members in this special unit are the central point for receiving, analyzing, and disseminating criminal intelligence information related to incendiary fires. All of the investigators and supervisors noted how beneficial the crime intelligence analysis unit and its personnel have been to their daily activities and the success of the Bureau.

Section 4. The Tennessee Interoperability Project--A Statewide, Multiagency Solution to Data and Intelligence Sharing

The leader in applying new technology to enhance data capture and intelligence sharing has been the Tennessee State Fire Marshal's Bomb and Arson Investigation Section (BAIS). Several years ago, a retired ATF official agreed to lead the office into the 21st century with better automation and technology for investigators. At that time, personnel were still using index cards and paper forms for bomb and arson incident reports. Computerized reporting was nonexistent. Neither was the office processing NIBRS reports through the FBI's UCR Index. Linking the headquarters office in Nashville with the field offices in Jackson and Knoxville was accomplished primarily by phone and fax machines.

For many years, USFA had worked on developing the concept and prototype of a computerized system that would move fire, arson, and bomb investigations into the electronic age with an information platform that could foster multiagency partnerships and provide a management tool for investigation unit managers who oversee and manage hundreds of cases. At the State level, case investigations easily can involve two or three different agencies, each with its own report forms and data files. Ideally, USFA's system design would employ commercial off-the-shelf software that then could be configured per specific user requirements. Those requirements would include handling complex information collection and information transfer using a variety of communications means (local and wide-area networks, virtual private networks, cell phones, satellite, and laptop computers), generally in a secure environment.

USFA established a working partnership with the Tennessee Valley Authority's (TVA) Police Department Special Projects Office and private-sector partners. Together they created a system derived in part from the Counterintelligence/Human Intelligence Management System Product Manager (previously used by the Army). Commercial off-the-shelf (COTS) software that was being used by some law enforcement agencies had a good combination of features and was added to the system. The final prototypes constituted a virtual office in which any number of field units could interact with the host server. The kits included a typical desktop suite; Global Positioning System (GPS) software and related peripherals; job-specific software; redundant connectivity capabilities and software; and digital photographic equipment.

By the year 2000, the system was ready to be tested. The development team needed to ascertain whether the system could serve as a viable statewide infrastructure and data repository with the potential to support both State and local investigation agencies. Specifically, the testing and evaluation would determine whether the model was capable of:

- servicing as a "force multiplier" by enhancing the quality and quantity of incident-related information and intelligence gathering;
- providing access to and collecting intelligence data and information during the active/ongoing investigation of an incident;
- providing a shared statewide repository for intelligence data (postevent) that could be used to mitigate crime; and
- preventing arson crimes from occurring.

The Tennessee State Fire Marshal's BAIS agreed to test the system prototype. Their involvement demanded considerable time and effort along with a director who could stay the course during the trial and error period for the system. For many reasons, the State of Tennessee was a good test bed. The State Fire Marshal's Office could connect to inputs from numerous other organizations and offices; the varied topography (river basin "flat lands," piedmont, and mountains) would test connectivity to the records system from different physical environments; and the State's computer center could host additional technology systems. In addition, the State had recently organized a mobile Special Operations Response Team (SORT) to handle major incidents. This team could demonstrate remote and mobile command and control using the new technology. Finally, BAIS's authority to investigate crimes of arson and criminal misuse of explosives throughout the State ensured access to all areas if the system could be expanded elsewhere in the State.

As BAIS investigators and managers began to work with the system, they developed a list of requirements and identified the constraints that had to be corrected. Their experience is valuable for any public safety agency that plans to establish a new or expanded information management and communications system with connectivity to other agencies.

Minimum Essential Requirements for Computerized Case Management and Intelligence Sharing System

1. Require minimal start-up time for the computer novice.

2. Provide a user-interface (UI) that can be customized to suit the agency's data and reporting needs.

3. Be able to function both in remote areas and in urban locations.

4. Operate as connected to the State's main intelligence/data system, or as a stand-alone field unit.

5. Provide technical support commensurate with the sophistication of the system.

6. Provide dual-functionality as an onscene, real-time crime-reporting tool, and as a repository of case-related data for case management, prosecutor support, intelligence sharing, and reporting.

7. Meet rigid security requirements.

8. Enable users to transcend and interrogate crime-based intelligence databases at all levels across jurisdictions.

9. Be compatible with existing computer hardware and software.

10. Be accessible to cleared users in other jurisdictions.

11. Generate a criminal case report along with exhibits.

12. Meet applicable laws.

Three key lessons emerged from the pilot testing of USFA's system. They were:

Lesson 1--Gather input from all groups that will be using the new system and itemize what features you **must** have, and what capabilities you **would like** to have. Itemize the list and distribute it for additional feedback. Then, communicate these requirements to the contractor installing the hardware and software, prior to testing.

Lesson 2--Before installing and using a system like Tennessee's, carefully analyze the operational environment (server, firewalls, telecommunications, and so forth), within which the new system will be operating. Make sure that the new system can be integrated successfully, and function as needed, taking into account connectivity requirements and security.

Lesson 3--A software engineer should test the configuration in the field to make sure that the machines are identical internally, that is, consistent in the way they operate and using the same priority order for using the software. The systems' configuration management is essential to ensure that there are no inherent software conflicts if other peripherals are added.

Through mid-2003, the Web-based system in Tennessee was used to great advantage in investigating 37 homicides, making 345 arrests, and gaining 220 convictions. By the close of fiscal year 2003, the State's percentage of arson cases cleared by arrest stood at 63 percent.

Applications for Emergency Onscene Reporting and Interagency Communications

A major earthquake hits with no warning. A bomb blast rips through a busy central business district. In both situations, hundreds of public safety workers from many agencies will be on the scene handling a myriad of response, recovery, and, in the second case, investigation duties. Key to everything will be communications--the basic ingredient of Incident Command, and control and coordination. And communications is the number-one factor that affects how well and how quickly lives are saved, people are rescued, and basic infrastructure is restored. Evaluations of Federal exercises, like Top Officers (TOPOFF) I and II and others, have shown that communications interoperability remains one of the biggest problems in a disaster environment.

USFA's system has the potential for addressing one of the many communications issues faced by first responders, as well as by their State and Federal disaster assistance partners. Immediate onscene images can be sent to cleared users whose computers are configured into the system. That means that a headquarters office, for example, could receive direct information from responders, IC's, and investigators. Resources could be channeled more efficiently to emergency sites and disaster zones and the information generated among the system participants would be available to all with appropriate access. (See Figure 5.)

Figure 5. Dual Use Web-Based System

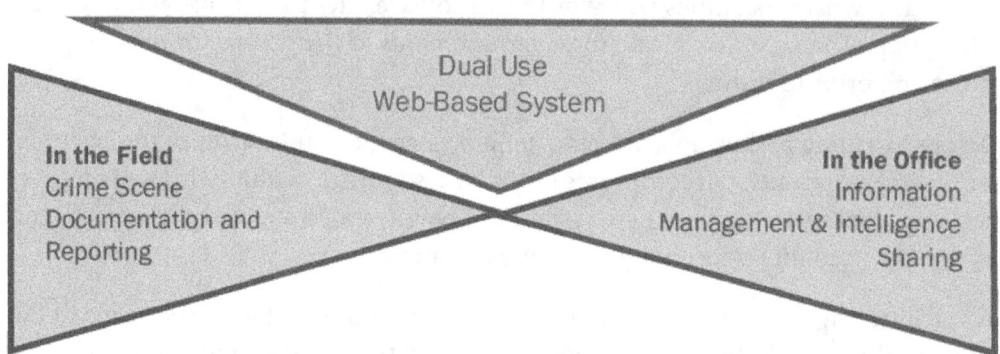

- real time date and images;
- onscene information and transmission;
- photos, maps, diagrams;
- rapid data exchange; and
- remote conferencing.

- repository of detailed incident data;
- master names indexes;
- modus operandi;
- case status;
- incident and summary reports;
- evidence logs;
- photos, maps, scene diagrams; and
- link to research sources.

Department of Homeland Security
Federal Emergency Management Agency

U.S. Fire Administration
Working for a Fire Safe America

FIRE/ARSON INVESTIGATION
UNIT TECHNICAL ASSISTANCE
PROJECT
DEADLINE DATE:

APPLICATION INSTRUCTIONS

Any jurisdiction wishing to apply for the technical assistance should complete the application form in as much detail as possible. Applicants may submit any additional material they think will be helpful.

Each application should also include a copy of the fire department's organizational chart, a staffing profile for the Fire Marshal's Office, and a description of the Unit's operational responsibility and authority.

Applications should be submitted and signed **ONLY** by individuals with supervisory authority over the Fire Investigation Unit to be reviewed. Please note, fire investigation units without police powers will require the concurrence of the participating law enforcement authority (as noted on Page 3) of the application form).

In order to be considered for the 2004 program year a complete application must be received by **February 27, 2004**. The application should be mailed or faxed to the following address:

> USFA-Arson/Fire Investigation Unit Technical Assistance Project
> *Joseph Ockershausen, Project Manager*
> TriData Corporation
> 1000 Wilson Boulevard, 30th Floor
> Arlington, VA 22209
> Phone: (703) 351-8300
> Fax: (703) 351-8383

For additional information, interested officials may contact either the Project Manager (identified above) or the USFA Program Manager Ken Kuntz, Fire Studies Specialist, at (301) 447-1271 or email: Ken.Kuntz@dhs.gov.

Department of Homeland Security
Federal Emergency Management Agency

U.S. Fire Administration
Working for a Fire Safe America

FIRE/ARSON INVESTIGATION
UNIT TECHNICAL
ASSISTANCE PROJECT
DEADLINE DATE:

SECTION 1 – BACKGROUND INFORMATION

1. REQUESTING ORGANIZATION

ADDRESS	Street:		City:	State:	Zip Code:
	Unit Phone:	Unit Fax:	Email Address:		

2. ARSON/FIRE INVESTIGATION COMMANDER (*Name/Title/Department*):

WORK PHONE:

3. TYPE OF JURISDICTION (*Please check one*):

❑ STATE ❑ CITY ❑ COUNTY ❑ OTHER

ESTIMATED POPULATION:

4. AGENCY CONDUCTING <u>FIRE ORIGIN & CAUSE</u>:

5. AGENCY CONDUCTING <u>CRIMINAL INVESTIGATIONS</u>:

A. NUMBER OF INVESTIGATORS ASSIGNED:
_____ FULL-TIME _____ PART-TIME

A. NUMBER OF INVESTIGATORS ASSIGNED:
_____ FULL-TIME _____ PART-TIME

B. ARE THEY CROSS-TRAINED IN LAW ENFORCEMENT?	❑ YES	❑ NO	B. ARE THEY CROSS-TRAINED IN FIRE INVESTIGATION?	❑ YES	❑ NO
C. POWER OF ARREST?	❑ YES	❑ NO	C. POWER OF ARREST?	❑ YES	❑ NO

6. IS THERE A SPECIFIC PROSECUTOR ASSIGNED TO OVERSEE ARSON CASES?	❑ YES	❑ NO

7. DOES YOUR JURISDICTION HAVE AN ACCELERANT DETECTION CANINE?   ❑ YES   ❑ NO

IF "NO", DO YOU BORROW ONE WHEN NEEDED?   ❑ YES   ❑ NO   FROM?_____

8. PLEASE CHECK ANY OF THE FOLLOWING WHICH APPLY TO YOUR INVESTIGATION AGENCY:

❑ National Fire Incident Reporting System (NFIRS) participant

❑ Investigators prepare draft reports using computers

❑ Active juvenile firesetter program

❑ Routinely utilize ATF support, as appropriate

❑ Use a case management system (AIMS) or equivalent

❑ Participate in UCR/NIBRS

❑ Investigators can access National Crime Information Center (NCIC)

❑ Investigation personnel also have EOD or Code Enforcement responsibilities.

❑ Formal inter agency fire investigation team (Please list agencies below)

_____

_____

9. WHAT ARE THE MOST PREVALENT MOTIVES FOR INCENDIARY FIRES IN YOUR JURISDICTION?

❑ CRIME CONCEALMENT	❑ VANDALISM	❑ FRAUD	❑ SPITE/REVENGE	❑ DOMESTIC/VIOLENCE	❑ OTHER

SECTION 2 – STATISTICAL OVERVIEW OF DEPARTMENT AND INVESTIGATIVE UNIT ACTIVITY

PLEASE PROVIDE THE DATA REQUESTED IN THE FOLLOWING CATEGORIES:

FIRES	2001	2002	INVESTIGATED FIRES	2001	2002
Total fire calls			Total investigation		
Total structure fires			Investigated structure fires		
Total vehicle fires			Investigated vehicle fires		

CAUSE DETERMINATION	Accidental	Incendiary	Undetermined
2001			
2002			

CASE OUTCOMES	2001	2002	JUVENILE DATA	2001	2002
Cases cleared by arrest			# of fires involving juveniles		
Cases accepted for prosecution			# of incendiary fires involving juveniles		
Total Convictions					
Adult			# of juveniles counseled for firesetting		
Juvenile					

SECTION 3 – ESTIMATED FISCAL AND HUMAN RESOURCE ALLOCATIONS

PLEASE IDENTIFY THE RESOURCES AVAILABLE TO YOUR UNIT

BUDGET ALLOCATED	2001	2002	NUMBER OF PERSONNEL	2001	2002
Total Fire Department			Total Fire Department		
• Fire Marshal's Office			• Fire Marshal's Office		
• Investigations			• Investigations		
• Inspections			• Inspections		
• Public Educations			• Public Educations		
• Overtime			• Overtime		

SECTION 4 – GENERAL INFORMATION

WHAT IS THE INVESTIGATORS' WORK SCHEDULE?

HOW ARSON IS IMPACTING YOUR JURISDICTION: (*please be as specific as possible*)

DESCRIBE THE WORKING RELATIONSHIP AMONG LAW ENFORCEMENT, FIRE, AND THE PROSECUTOR'S OFFICE: (*please be as specific as possible*)

WHAT SPECIFIC PROBLEMS ARE YOU FACING WITHIN YOUR UNIT? (*please be as specific as possible*)

SECTION 5 – LOCAL POINTS OF CONTACT

TITLE	NAME	PHONE	EMAIL
FIRE CHIEF			
FIRE MARSHAL			
POLICE CHIEF/SHERIFF			
PROSECUTOR/DISTRICT ATTORNEY			
ATF CONTACT			

SECTION 6 – AUTHORIZATION TO APPLY

*IF YOUR AGENCY RELIES ON ANOTHER AGENCY FOR CRIMINAL INVESTIGATION SERVICES, THAT AGENCY MUST AGREE TO PARTICIPATE IN THE PROJECT ALSO. (*Please indicate this below.*)

REQUESTING ORGANIZATION:	
NAME/TITLE OF CHIEF OFFICIAL:	
WORK PHONE:	
EMAIL:	
SIGNATURE	
DATE:	

*COOPERATING ORGANIZATION:	
NAME/TITLE OF CHIEF OFFICIAL:	
WORK PHONE:	
EMAIL:	
SIGNATURE	
DATE:	

Appendix B
List of All Participating Jurisdictions

USFA FIRE AND ARSON INVESTIGATION UNIT TECHNICAL ASSISTANCE PROJECT 1989-2004	
TYPE OF AGENCY	YEAR OF STUDY
STATE	
1. Nevada State Arson Unit	1991
2. Tennessee State Fire Investigation Unit	1991
3. Minnesota State Fire Investigation Unit	1993
4. New Hampshire State Fire Investigation Unit	1993
5. Kansas State Fire Investigation Unit	1994
6. Maine State Fire Investigation Unit	1994
7. California State Fire Investigation Unit	1995
8. Indiana State Fire Investigation Unit	1995
9. Missouri State Fire Investigation Unit	1995
10. Vermont State Police	1995
11. Delaware State Fire Marshal's Office	1996
12. Texas Commission on Fire Protection	1996
13. Utah State Fire Marshal	1997
14. Oklahoma State Fire Marshal's Office	1998
15. Alaska State Fire Marshal's Office	1999
16. Maryland State Fire Marshal's Office	1999
17. Massachusetts State Fire Marshal's Office	1999
18. Rhode Island State Fire Marshal's Office	1999
19. Illinois Office State Fire Marshal	2001
20. Ohio Division of State Fire Marshal	2002
21. Wyoming State Fire Marshal's Office	2002
22. Louisiana State Fire Marshal's Office	2003
23. Florida State Fire Marshal's Office	2003
24. Iowa State Fire Marshal's Office	2004
COUNTY	
25. Kitsap County Fire Investigation Agencies, WA	1989
26. Livingston County Fire Department, MI	1989
27. Baltimore County Fire Investigation Division, MD	1991
28. Humbolt County Fire/Arson Investigation Unit, CA	1991
29. Imperial County Fire Department, CA	1991
30. Mohave County Fire Investigation Task Force, AZ	1991
31. Montgomery County Fire and Rescue Service, MD	1993
32. King County Fire Marshal's Office, Fire Investigation Unit, WA	1994
33. Palm Beach County Fire/Rescue Investigations Section, FL	1994
34. Cobb County Fire and Rescue, GA	1996
35. Henrico County Division of Fire, VA	1997
36. Orange County Fire Department, CA	1997
37. Prince George's County Fire Department, MD	1997
38. Lexington Division of Fire, KY	1998
39. Clayton County Fire Department, GA	1999

USFA FIRE AND ARSON INVESTIGATION UNIT TECHNICAL ASSISTANCE PROJECT 1989-2004	
40. Fulton County Fire Department, GA	1999
41. Los Angeles County Fire Department, CA	1999
42. District Attorney's Office of Chester County, PA	2000
43. Fairfax County Fire and Rescue, VA	2001
44. Knox County Fire Investigation, TN	2001
45. Anne Arundel County Fire Investigation, MD	2001
46. Harris County Fire and Emergency Services Department, TX	2001
47. Maricopa County Arson Task Force, AZ	2002
48. Pierce County Fire Prevention Bureau, WA	2002
49. Santa Barbara County Fire Department, CA	2002
50. Wake County Public Safety Fire and Rescue Service Division, NC	2002
51. Snohomish County Fire Marshal Office, WA	2003
52. Gwinnett County Dept. of Fire and EMS, GA	2004
53. Hillsborough County Fire Rescue, FL	2004
CITY	
54. Gainesville Fire Investigation Unit, FL	1989
55. Norfolk Fire Investigation Unit, VA	1989
56. Orlando Fire Department, FL	1989
57. Wilmington Fire Department, NC	1989
58. Asheville Fire Department, NC	1990
59. Aurora Fire Department, CO	1991
60. Columbia Fire Department, SC	1991
61. City of Garland, Fire Investigation Unit, TX	1991
62. New Jersey Fire Investigation Unit, Pleasantville, NJ	1991
63. District of Columbia Fire Investigation Division, Washington, DC	1991
64. Lawrence Arson Task Force, MA	1992
65. Albuquerque Fire Department, NM	1993
66. Bridgeport Fire Investigation Division, CT	1993
67. City of Buffalo Fire Investigation Unit, NY	1993
68. Butte Silver Bow Fire Investigation Unit, MT	1993
69. Indianapolis Fire Department, IN	1993
70. Omaha Fire Investigation Bureau, NE	1993
71. Fresno Fire Department, Fire Investigations Unit, CA	1994
72. Memphis Fire Services, Fire Investigations Unit, TN	1994
73. Oklahoma City Fire Department, OK	1994
74. Saginaw Fire Department, Fire Investigations Unit, MI	1994
75. The Metropolitan Fire Investigation Unit, Youngstown, OH	1994
76. Birmingham Fire and Rescue Service Department, AL	1995
77. Louisville Division of Fire Arson Investigation Bureau, KY	1995
78. New Orleans, LA	1995
79. Philadelphia Fire Department, Fire Investigation Unit, PA	1995
80. Richmond Fire Department, VA	1995
81. Rochester Fire Department, NY	1995
82. San Francisco Fire Department, Arson Squad, CA	1995
83. Charlotte Fire Investigation Task Force, NC	1996
84. City of Chicago Fire Department, IL	1996

USFA FIRE AND ARSON INVESTIGATION UNIT TECHNICAL ASSISTANCE PROJECT 1989-2004	
85. Elizabeth Fire Department, NJ	1996
86. Houston Fire Department, TX	1996
87. Portland Fire Bureau Fire and Arson, OR	1996
88. Tucson Fire Department, AZ	1996
89. Worcester Fire Investigation Unit, MA	1996
90. Austin Fire Department, TX	1997
91. Columbus Division of Fire, OH	1997
92. Henrico Metro Force, VA	1997
93. Minneapolis Fire Department, MN	1997
94. City of Oakland Fire Services Agency, CA	1997
95. Virginia Beach Fire Department, VA	1997
96. Cleveland Fire Investigation Unit, OH	1998
97. Des Moines Fire Department, IA	1998
98. Flint Fire Department, MI	1998
99. Honolulu Fire Department/Honolulu Police Department, HI	1998
100. Kansas City Fire Department, KS	1998
101. Mesa Fire Prevention Bureau, AZ	1998
102. San Antonio Fire Department, TX	1998
103. Alpharetta Fire Department, GA	1999
104. Atlanta Fire Department, GA	1999
105. College Park Fire Department, GA	1999
106. Fairbanks Department of Public Safety, AK	1999
107. Bakersfield Fire Department, CA	2000
108. Colorado Springs Fire Department, CO	2000
109. District of Columbia Fire/EMS Department, DC	2000
110. El Paso Fire Department, TX	2000
111. City of Fort Lauderdale Fire-Rescue Department, FL	2000
112. Salt Lake City Fire Department, UT	2000
113. Savannah Fire and Emergency Services, GA	2000
114. Toledo Fire and Rescue Department, OH	2000
115. The Wichita Fire Department, KS	2000
116. Phoenix Fire Department, AZ	2000
117. Baltimore City Fire & Police Department, MD	2001
118. Decatur Fire Department, IL	2001
119. Denver Fire Department, CO	2001
120. Ft. Worth Fire Department, TX	2001
121. Hampton Division of Fire and Rescue, VA	2001
122. Lansing Fire Department, MI	2001
123. Paterson Fire Department, NJ	2001
124. Mobile Fire-Rescue Department, AL	2002
125. Tacoma Fire Department, WA	2002
126. Duluth Fire Department, MN	2002
127. City of Rochester Fire Department, NY	2002
128. Long Beach Fire Department, CA	2002
129. Portsmouth Fire and Rescue, VA	2002
130. San Diego Fire and Life Safety Service, CA	2002

USFA FIRE AND ARSON INVESTIGATION UNIT TECHNICAL ASSISTANCE PROJECT 1989-2004	
131. Tulsa Fire Department, OK	2002
132. Ft. Wayne Fire Department	2003
133. Stockton Fire Department, CA	2003
134. Dayton Fire Department, OH	2003
135. Memphis Fire Protection Bureau, TN	2003
136. Gastonia Fire Department, NC	2003
137. City of Seattle Fire Department, WA	2003
138. Louisville Fire and Rescue, KY	2003
139. Louisville Metro Criminal Justice, KY	2003
140. San Francisco Fire Department, CA	2004
141. Nashville Fire Department, TN	2004
142. Sacramento Metropolitan Fire District, CA	2004
143. Tampa Fire Rescue-Fire Prevention Bureau, FL	2004

Appendix C
Copy of Ohio's Seminar Program Relative to Prosecution of Arson Cases

[§ 3737.33.1] § 3737.331. Seminar program relative to prosecution of arson cases.

The fire marshal, after consultation with prosecuting attorneys of this state selected with due regard for geographic, urban, and rural representation, shall make available a seminar program, attendance at which is optional, that is designed to provide the prosecuting attorney and an assistant prosecuting attorney from each county of this state with current information, data, training, and techniques relative to the prosecution of arson cases. The fire marshal shall cooperate with the attorney general in the establishment of the seminar program. The fire marshal shall offer the seminar program at least twice annually.

Each prosecuting attorney may personally attend, or may require an assistant prosecuting attorney to attend, one of the seminar programs annually. While attending a seminar program offered by the fire marshal, each prosecuting or assistant prosecuting attorney shall receive his full regular compensation from the county by which he is employed.

Appendix D
Sample of Work Schedule from Project Jurisdictions

Fire Investigation Work Schedules 1

Alaska State Fire Marshal's Office 1

Delaware Alaska State Fire Marshal's Office 1

Maine Alaska State Fire Marshal's Office 1

Maryland Alaska State Fire Marshal's Office 1

Massachusetts Alaska State Fire Marshal's Office 2

Rhode Island Alaska State Fire Marshal's Office 2

Utah Alaska State Fire Marshal's Office 3

Fairbanks, Alaska 3

Mesa, Arizona 3

Tucson, Arizona 3

Bakersfield, California 3

Fresno, California 4

Ft. Lauderdale, Florida 4

West Palm Beach Co., Florida 5

Cobb County, Georgia 5

Fulton Co., Georgia 5

Des Moines, Iowa 6

Honolulu, Hawaii 6

Kansas City, Kansas 6

Flint, Michigan 6

Oklahoma, Oklahoma 7

El Paso, Texas 7

San Antonio, Texas 7

Salt Lake City, Utah 8

Virginia Beach, Virginia 8

FIRE INVESTIGATION WORK SCHEDULES

Alaska State Fire Marshals Office

The normal shift for sworn personnel is straight day work, Monday through Friday. After-hours call-out for fire investigation is handled in each regional office.

Delaware State Fire Marshal's Office

The New Castle office has the highest caseload and, as a result, has the greatest number of staff. Moreover, coverage in that county is 16 hours per day. This allows for a day shift of five deputies, and a later shift covered by one deputy working 3 p.m. to 11 p.m., Monday through Friday. The other hours between 11 p.m. and 8 a.m. are covered by callback personnel on a 3-day rotation.

Investigators at the Sussex and Kent County offices work day shifts with callback rotation for evenings, nights, and weekends. Sussex County offices uses an on-call rotation of three consecutive 24-hour days. The Kent County office uses a seven consecutive, 24-hour day callback rotation. Kent County has a less demanding caseload than does Sussex County.

Maine State Fire Marshal's Office

The field investigators regularly work a daytime shift and use an on-call system to cover responses that occur after normal duty hours. Due to a union contract provision, there is little flexibility permitted in scheduling.

Each investigator submits a daily activity log that describes his or her daily activities in a general manner. These are reviewed by the respective northern and southern area supervisors.

Maryland State Fire Marshal's Office

The normal work shift for sworn personnel is straight day work, Monday through Friday, with the options being available to the Regional Supervisor for late work. An "on-call" Deputy State Fire Marshal is available in each region to respond to emergency calls after normal working hours and on holidays and weekends.

Massachusetts State Fire Marshal's Office

Personnel work a standard administrative schedule from 8 a.m. to 5 p.m. Monday through Friday, and are off on weekends and holidays. During all other periods, investigation personnel are on call. Each region establishes call-out procedures based on the specific needs for that region. Call-outs can vary; they are normally conducted on a rotational basis depending on the regional staffing and assignment. Some investigators, such as those detailed to canine units, are always on call, and are dispatched whenever needed. Others investigators are on call Monday through Friday and are responsible for their assigned region.

Rhode Island State Fire Marshal's Office

State fire investigators work an administrative schedule from 8:30 a.m. to 4:30 p.m. Monday through Friday. After hours, weekends, and holidays are covered on a call-out basis. The State Fire Marshal's Office has developed a formal procedure outlining the steps necessary to request the on-call investigator. Fire investigators are subject to call-outs between the hours of 4:30 p.m. and 8:30 a.m. One investigator is normally scheduled on call from 8:30 a.m. Monday until the following Monday morning at 8:30 a.m. All after-hour requests for fire investigators are directed through the

Department of Environmental Management Enforcement dispatcher who alerts the on-call investigator. The on-call fire investigator is expected to arrive on the emergency scene within 1 hour from the initial notification.

Each fire investigator is assigned a specific geographical area of responsibility, and is responsible for the investigation and followup of all fires occurring within their assigned area. The on-call investigator is responsible for covering any fire within the State that occurs after normal work hours, on weekends, and on holidays. Should the on-call investigator respond to a fire within his or her assigned area, he or she will retain possession of the fire. If the fire occurs outside of his/her assigned area of responsibility, the on-call investigator will conduct the initial interviews and fire scene examination. After the investigation, he/she forwards the findings to the appropriate fire investigator for followup. Each investigator is responsible for coordinating his/her fire investigation efforts with the local fire and law enforcement officials, and the State Attorney General's Office.

Utah State Fire Marshal's Office

The investigators regularly work a daytime shift (Monday through Friday, 7:30 a.m. to 4 p.m.) with some flexible start/finish times. A pager system is used to cover responses that are required after normal duty hours.

There is no compensation for standby time and there is no overtime compensation provided for response outside normal business hours. The investigators accumulate credit for time off when they respond to incidents after business hours and on weekends and holidays.

Fairbanks, Alaska

The Deputy Fire Marshal's normal shift is straight day work, Monday through Friday. He/She is subject to after-hours callback, as necessary, to fulfill responsibilities in both code enforcement and fire investigation. Battalion Chiefs also are called back to work as necessary to conduct fire investigations.

Mesa, Arizona

Employees assigned to the unit work a rotating shift, shifts A, B, and C. The daily shift is during normal business hours, Monday through Friday. Responses after hours are handled on a callback basis. Overtime is paid for callback. Each investigator is assigned a shift (A, B, or C).

Tucson, Arizona

The fire investigation captain and the three investigators maintain a regular work schedule of 8 a.m. to 5 p.m., Monday through Friday, excluding holidays. There is a rotation of the three investigators into an on-call status for fire investigations who may be required outside regular business hours. The investigators are oncall every third week for 7 consecutive nights at a time. All investigations conducted outside normal business hours are on an overtime basis.

The Captain may respond to major incidents outside normal business hours.

Bakersfield, California

The three captain/investigators work a standard administrative work schedule Monday through Friday with some flexibility in the starting time. The investigators respond to calls from the field in accordance with Department Policy #404. This policy defines the specific instances to request an investigator, and those situations not requiring their presence. Generally, investigators respond to

- all structure and vehicle fires determined to be incendiary;
- all structure and vehicle fires where the cause cannot be determined;
- all major grass fires determined to be incendiary, and those where the cause cannot be readily determined;
- all burns resulting in death or hospitalization.

After hours, weekends, and holidays are covered by the stand-by investigator on a call-out basis. Fire investigation personnel are assigned to standby duty on a biweekly rotational basis among the three investigator captains. The duty captain will take all calls on a 24-hour basis, unless that investigator is on leave or the incident occurs during the normal work schedule, such as a flex-day. An alternate captain will cover the duty captain's assigned coverage period if the duty captain is scheduled to be off. If the duty captain is already committed to an incident, the alternate captain will be alerted to any subsequent call-outs. Additional investigators can be called out if more than one investigator is needed to conduct an investigation.

Standard operational procedures (SOP's) require fire suppression units to maintain custody of the fire scene until the arrival of the fire investigator.

Fresno, California

The supervisor and the investigators rotate on call in order to provide response to incidents outside normal business hours. This type of response is on an overtime basis, at time and a half, with a minimum of 2 hours for each required response.

Ft. Lauderdale, Florida

Fire department investigators work a standard 40-hour administrative work schedule, with evenings, weekends, and holidays off. Investigators have the option of either working four 10-hour days or five 8-hour days. Currently, all but one fire department investigator has chosen to work the four 10-hour days, taking Fridays off. The remaining investigator has chosen to work a standard 5-day workweek, thus eliminating the potential for call-outs on Fridays. Personnel who are assigned to day-work positions receive a premium pay differential. This is an attractive perk used as an incentive to retain experienced personnel on day work. After-hour emergencies, weekends, and holidays are covered through a formal call-out procedure established by the fire department. Personnel are on call for 7 consecutive days. On-call periods cover the 24 hour period beginning at 8 a.m. Tuesday morning and concluding the following Tuesday at 8 a.m. If additional investigators are needed, the prior week's on-call investigator is summoned.

Police investigators also work an administrative schedule consisting of 8-hour days, Monday through Friday, with staggered starting times. After-hour emergencies during evenings, weekend, and holidays periods are handled on a call-out basis. Call-out periods rotate between the police investigators on a seven-day cycle. Police investigators are alerted through the Fire Department Communications Division.

West Palm Beach County, Florida

The schedule now used by the Fire Investigation Section as investigators working two shifts--one beginning at 7:30 a.m. and finishing at 4 p.m., the second beginning at 3 p.m. and ending at 11:30 p.m.--Monday through Friday. On Saturday and Sunday, a third shift works 3 p.m.-11:30 p.m. All other times and on holidays, one person is assigned to callback, which requires overtime compensation for investigators.

Cobb County, Georgia

The division is staffed with three investigators who are on duty Monday through Friday. Evenings and weekends are covered by an investigator on call for overtime compensation. The on-call duty is rotated on a weekly basis. Investigators are available 24 hours a day, 7 days a week by scheduled work day or callback. The investigator can be alerted by radio pager or telephone. Investigators also have cell phones in unmarked responses vehicles.

Fulton County, Georgia

Personnel assigned to the Fulton County fire investigation unit work (Monday through Friday) 8 a.m. to 5 p.m. Emergencies after hours, on the weekend, and on holidays are covered through a formal call-out procedure established by the fire department. The State of Georgia is a right-to-work State. Supervisors often adjust an employee's hours accordingly or award hour-for-hour compensatory time in remuneration for those hours worked by personnel more than their normal shift. Overtime is very seldom an option. This practice should be reviewed. Management should take a cautious approach to arbitrarily adjusting employee work hours; this could be viewed as a violation of the Fair Labor Standards Act (FLSA).

Des Moines, Iowa

The Des Moines FIU operates a two-person team approach. One Senior Fire Inspector and one Police Detective make up each team. They work a 40-hour schedule, 5 days a week, 8 hours per day. Each investigative team rotates on-call procedures for after-hour calls. The on-call rotation runs a full week. All investigations conducted outside normal business hours are considered overtime. Investigators must log all overtime used and are given permission to use overtime, as necessary, to complete their investigations.

Honolulu, Hawaii

The investigations unit has a captain and two firefighter III investigators assigned. All three of the investigators are on an 8-hour day work schedule. The evenings and nights are covered by callback on a 7-day straight rotation. Currently, because the written report backlog has become so significant, one investigator has been reassigned to complete written reports, thereby reducing the section's complement to two. The Honolulu Police Department (HPD) detective sergeant also works straight day work. This detective responds to after-hour investigations when requested on a callback basis. The detective also is expected to cover a part of the general assignment caseload rotation. The HPD in the past had assigned two detectives to share the criminal fire case followup. Since the project sight visit, a second detective has been selected and is being scheduled for training.

Kansas City, Kansas

All personnel are assigned straight day work. The investigator is paged or telephoned to respond to incidents after normal working hours.

Flint, Michigan

The police detective sergeants work a day schedule of 10 days in a row with 4 days off. This shift provides weekend coverage; however, on Sundays they have primary assignment for domestic violence cases. The Fire Department's Fire Prevention Division FUI has five lieutenant inspector/investigators assigned and one captain supervisor. The captain supervises the entire fire prevention effort including inspections, public education, juvenile firesetting program, and the Arson Unit. All the fire prevention personnel work an 8-hour day, Monday through Friday. The deputy chief/fire marshal is the overall fire prevention commander.

Oklahoma City, Oklahoma

The agents regularly work a daytime shift (Monday through Friday) with some flexible start/finish times. An on-call system is used to cover responses that are required after normal duty hours.

El Paso, Texas

Uniformed personnel assigned to the Fire Investigation Section work an administrative schedule 8 a.m. to 5 p.m. (Monday through Friday), with an uncompensated hour for lunch. After hours, weekends, and holidays are covered on a call-out basis.

San Antonio, Texas

Employees assigned to the Arson Bureau work the following regular hours, with the exception of the captain and lieutenant, who work a 40-hour, 5-day work week. The said work week consists of two shifts consisting of the day shift and evening shift. The shifts are broken down as follows:

	Sun	Mon	Tues	Wed	Thurs	Fri	Sat
7 a.m. - 5 p.m.			A	A	A	A	
			B	B	B	B	
		C	C	C	C		
		D	D	D	D		
4 p.m. - 2 a.m.	E	E	E	E			
				F	F	F	F

Each letter represents a team of two investigators working a 10-hour shift

- 5 -

Salt Lake City, Utah

The investigators work a normal daytime Monday through Friday work week with flexible starting and ending times. Response to fire scenes after normal business hours, on weekends and holidays are covered on a rotating call-out schedule. Each investigator is "on call" for one out of every four weeks. Investigators who are "on call" are compensated with 1 hour of straight time per day. Compensation for a "callback" is a minimum of 4 hours. An agreement has been reached between the city and Salt Lake County Fire Department where a Salt Lake County investigator covers the "on call" every fourth week. This arrangement results from the fact that the County Fire Department's investigation unit is financed through the County's general fund and is available to all jurisdictions in Salt Lake County.

The fire department has issued a written policy entitled, "Callback and Notification Policy," which addresses procedures to be followed when calling back various personnel, including fire investigators. This policy also addresses who needs to be notified and what types of personnel should be called backed to duty, depending on the specific type of incidents.

Virginia Beach, Virginia

Monday through Thursday three investigators work dayshift, one investigator works 2 p.m. to 10 p.m. On a rotational basis, a different investigator also works 2 p.m. to 10 p.m. to provide two investigators on duty for the late shift. Fridays everyone works day work with the latest duty shift between 10 a.m. to 6 p.m. Fridays after 6 p.m. until Monday mornings are covered by rotational callbacks. Callout after 10 p.m. Monday through Thursday is handled by the rotational 2 to 10 p.m. evening shift investigator. The advantage of this shift is that the staffing is at full complement between 2 p.m. and 4 p.m. This allows for face-to-face communication among investigators and is a good period of time to complete followup interviews. This shift is conducive for followup and continuity on open cases. The fire department provides good compensation for callbacks. The department pays 1 hour for every 4 hours of weekend standby and time and a half for actual call outs.

www.ingramcontent.com/pod-product-compliance
Lightning Source LLC
Chambersburg PA
CBHW081219170526
45165CB00009B/2873